Nanophotonics and Nanofabrication

Edited by
Motoichi Ohtsu

Further Reading

Wehrspohn, R. B., Kitzerow, H.-S., Busch, K. (eds.)

Nanophotonic Materials

Photonic Crystals, Plasmonics, and Metamaterials

2008
Hardcover
ISBN: 978-3-527-40858-0

Maldovan, M., Tomas, E. L.

Periodic Materials and Interference Lithography

for Photonics, Phononics and Mechanics

2008
Hardcover
ISBN: 978-3-527-31999-2

Vollath, D.

Nanomaterials

An Introduction to Synthesis, Properties and Applications

2008
Hardcover
ISBN: 978-3-527-31531-4

Ostrikov, K., Xu, S.

Plasma-Aided Nanofabrication

From Plasma Sources to Nanoassembly

2007
Hardcover
ISBN: 978-3-527-40633-3

Rao, C.N.R., Müller, A., Cheetham, A. K. (eds.)

Nanomaterials Chemistry

Recent Developments and New Directions

2007
Hardcover
ISBN: 978-3-527-31664-9

Wolf, E. L.

Nanophysics and Nanotechnology

An Introduction to Modern Concepts in Nanoscience

Second, Updated and Enlarged Edition
2006
Softcover
ISBN: 978-3-527-40651-7

Misawa, H., Juodkazis, S. (eds.)

3D Laser Microfabrication

Principles and Applications

2006
Hardcover
ISBN: 978-3-527-31055-5

Nanophotonics and Nanofabrication

Edited by
Motoichi Ohtsu

WILEY-
VCH

WILEY-VCH Verlag GmbH & Co. KGaA

The Editor

Prof. Motoichi Ohtsu
University of Tokyo
Dept. of Electronics Engin.
2-11-16 Yayoi, Bunkyo-ku
Tokyo 113-8656
Japan

All books published by **Wiley-VCH** are carefully produced. Nevertheless, authors, editors, and publisher do not warrant the information contained in these books, including this book, to be free of errors. Readers are advised to keep in mind that statements, data, illustrations, procedural details or other items may inadvertently be inaccurate.

Library of Congress Card No.: applied for

British Library Cataloguing-in-Publication Data
A catalogue record for this book is available from the British Library.

Bibliographic information published by the Deutsche Nationalbibliothek
Die Deutsche Nationalbibliothek lists this publication in the Deutsche Nationalbibliografie; detailed bibliographic data are available on the Internet at http://dnb.d-nb.de.

© 2009 WILEY-VCH Verlag GmbH & Co. KGaA, Weinheim

All rights reserved (including those of translation into other languages). No part of this book may be reproduced in any form – by photoprinting, microfilm, or any other means – nor transmitted or translated into a machine language without written permission from the publishers. Registered names, trademarks, etc. used in this book, even when not specifically marked as such, are not to be considered unprotected by law.

Typesetting Thomson Digital, Noida, India
Printing betz-druck GmbH, Darmstadt
Binding Litges & Dopf GmbH, Heppenheim
Cover Design Adam-Design, Weinheim

Printed in the Federal Republic of Germany
Printed on acid-free paper

ISBN: 978-3-527-32121-6

Contents

Preface *XI*
List of Contributors *XV*

1   **Introduction** *1*
    Motoichi Ohtsu
1.1   History *1*
1.2   Fiber Probes and Sensing Systems *2*
1.3   Theory *3*
1.4   Devices *6*
1.5   Fabrications *9*
1.6   Applications to Systems and Evolution to Related Sciences *11*
1.7   Toward the Future *12*
    References *13*

2   **Nanofabrication Principles and Practice** *17*
    Tadashi Kawazoe and Motoichi Ohtsu
2.1   Adiabatic Nanofabrication *17*
2.2   Nonadiabatic Nanofabrication *19*
2.2.1   Nonadiabatic Near-Field Optical Chemical Vapor Deposition *19*
2.2.2   Nonadiabatic Near-Field Photolithography *27*
    References *33*

3   **Nanofabrications by Self-Organization and Other Related Technologies** *35*
    Takashi Yatsui, Wataru Nomura, Kazuya Hirata, Yoshinori Tabata, and Motoichi Ohtsu
3.1   Introduction *35*
3.2   Near-Field Optical Chemical Vapor Deposition *35*
3.3   Self-Assembling Method Via Optical Near-Field Interactions *41*
3.3.1   Regulating the Size and Position of Nanoparticles Using Size-Dependent Resonance *41*

3.3.2	Self-Assembly of Nanoparticles Using Near-Field Desorption *45*
3.3.3	One-Dimensional Alignment of Nanoparticles Using an Optical Near-field *49*
3.4	Near-Field Imprint Lithography *55*
3.5	Nonadiabatic Optical Near-Field Etching *60*
	References *64*
4	**Fabrication of Quantum Dots for Nanophotonic Devices** *69*
	Kouichi Akahane and Naokatsu Yamamoto
4.1	Introduction *69*
4.2	Fabrication of Self-Assembled QDs *71*
4.2.1	Control of Density and Emission Wavelength of Self-Assembled QDs *73*
4.2.2	Shortening Emission Wavelengths of Self-Assembled QDs *73*
4.2.3	Controlling the Density of Self-Assembled QDs *77*
4.2.4	Fabrication of Self-Assembled QDs with Antimonide-Related Materials *80*
4.2.5	Fabrication of Ultrahigh-Density QDs *83*
4.2.6	Summary *90*
4.3	Fabrication Techniques of Site-Controlled Nanostructures *91*
4.3.1	Nanopositioning Technique for Quantum Structures with Dioxide Mask *91*
4.3.2	Artificially Prepared Nanoholes for Arrayed QD Structure Fabrication *95*
4.3.3	Nanojet Probe Method for Site-Controlled InAs QD Structure *96*
4.3.4	Scanning Tunneling Probe Assisted Nanolithography for Site-Controlled Individual InAs QD Structure *98*
4.3.5	Metal-Mask MBE Technique for Selective-Area QD Growth *99*
4.4	Silicon-Related Quantum Structure Fabrication Technology *100*
4.4.1	III-V Compound Semiconductor QD on a Si Substrate *100*
4.4.2	Fabrication Technique of Silicon Nanoparticles as Si-QD Structures *101*
	References *103*
5	**ZnO Nanorod Heterostructures for Nanophotonic Device Applications** *105*
	Gyu-Chul Yi
5.1	Introduction *105*
5.2	ZnO Axial Nanorod Quantum Structures *108*
5.3	ZnO Radial Nanorod Heterostructures *117*
5.4	Conclusions *126*
	References *128*

6		**Lithography by Nanophotonics** 131
		Ryo Kuroda, Yasuhisa Inao, Shinji Nakazato, Toshiki Ito, Takako Yamaguchi, Tomohiro Yamada, Akira Terao, and Natsuhiko Mizutani
6.1		Introduction 131
6.2		Principle of the Optical Near-Field Lithography 132
6.3		Optical Near-Field Lithography System 136
6.4		Fabricated Patterns by Optical Near-Field Lithography 140
6.5		Improvement of Resolution and Fabricated Ultrafine Patterns 142
6.6		Summary 144
		References 146
7		**Nanopatterned Media for High-Density Storage** 147
		Hiroyuki Hieda
7.1		Introduction 147
7.2		Nanopatterned Media 149
7.2.1		Fabrication Process of Nanopatterned Media 150
7.3		Block-Copolymer Lithography for Nanopatterned Media 153
7.3.1		Self-Assembled Phase Separation of Block-Copolymers 153
7.3.2		Fabrication of Magnetic Nanodots by Block-Copolymer Lithography 155
7.4		Control of Orientation of Self-Assembled Periodic Patterns of Block-Copolymers 158
7.5		Summary 162
		References 164
8		**Nanophotonics Recording Device for High-Density Storage** 167
		Tetsuya Nishida, Takuya Matsumoto, and Fumiko Akagi
8.1		Introduction 167
8.2		Thermally Assisted Magnetic Recording Simulation 168
8.3		The 'Nanobeak,' a Near-Field Optical Probe 170
8.4		Bit-Patterned Medium with Magnetic Nanodots 172
8.5		Hybrid Recording Experiment 174
8.6		Near-Field Optical Efficiency in Hybrid Recording 175
8.7		Summary 177
		References 178
9		**X-ray Devices and the Possibility of Applying Nanophotonics** 179
		Masato Koike, Shinji Miyauchi, Kazuo Sano, and Takashi Imazono
9.1		Introduction 179
9.2		Design of the Multilayer Laminar-Type Grating 180
9.3		Specification of the Multilayer Laminar-Type Grating 183
9.4		Fabrication of Multilayer Laminar-Type Gratings 183
9.5		Simulation of Diffraction Efficiency 185
9.6		Measurement of Diffraction Efficiency 186
9.7		Roughness Evaluation using Debye–Waller Factors 189
		References 190

10		**Nanostructuring of Thin-Film Surfaces in Femtosecond Laser Ablation** *193*
		Kenzo Miyazaki
10.1		Introduction *193*
10.2		Experimental *194*
10.3		Properties of Nanostructuring *194*
10.3.1		Polarization *195*
10.3.2		Multiple Pulses *195*
10.3.3		Fluence *195*
10.3.4		Laser Wavelength *197*
10.3.5		Pulse Width *197*
10.4		Bonding-Structure Change *197*
10.5		Dynamic Processes *198*
10.5.1		Reflectivity of Ablating Surface *199*
10.5.2		Ultrafast Dynamics *202*
10.6		Local Fields *204*
10.7		Origin of Periodicity *208*
10.8		Summary *212*
		References *212*

11		**Quantum Dot Nanophotonic Waveguides** *215*
		Lih Y. Lin and Chia-Jean Wang
11.1		Conceptual Formation and Modeling of the Device *216*
11.1.1		QD Gain vs. Pump Power *218*
11.1.2		FDTD Modeling for Interdot Coupling *220*
11.1.3		Monte Carlo Simulation for Transmission Efficiency *221*
11.2		From Concept to Realization – Fabrication of the Device *224*
11.2.1		DNA-Directed Self-Assembly Fabrication *224*
11.2.1.1		Self-Assembly Process and Characterization *224*
11.2.1.2		Programmable DNA-Directed Self-Assembly *227*
11.2.2		Self-Assembly Through APTES *228*
11.2.2.1		Self-Assembly Process and Characterization *228*
11.2.2.2		Multiple-QD-Type Waveguide Fabrication *231*
11.2.3		Discussion on Fabrication Methods *232*
11.3		How Well the Devices Work – A First Probe *233*
11.3.1		Waveguiding with Flexibility *234*
11.3.2		Loss Characterization *235*
11.4		To Probe Further – Summary and Outlook *236*
		References *238*

12		**Hierarchy in Optical Near-fields and its Application to Nanofabrication** *241*
		Makoto Naruse, Takashi Yatsui, Hirokazu Hori, Kokoro Kitamura, and Motoichi Ohtsu
12.1		Introduction *241*

12.2	Angular Spectrum Representation of Optical Near-Fields	*242*
12.3	Generation of Smaller-Scale Structures via Optical Near-Fields: A Theoretical Basis	*244*
12.4	Experiment	*247*
12.5	Conclusion	*249*
	References	*250*

Index *253*

Preface

This book outlines the principles and practices of nanofabrication based on the novel optical technology of *nanophotonics*, which utilizes the optical near-field, i.e. the nanometer-sized light that is localized on the surface of a nanometric material. In the early 1980s, the editor of this book (M. Ohtsu) started his pioneering research on optical near-fields because he judged that optical near-fields would be required to change the paradigm underlying optical science and technology. In the 1990s, conventional optical technology progressed very rapidly and the photonics industry developed, but further progress became difficult due to a fundamental limit of light known as the diffraction limit. However, there was a growing awareness among scientists and engineers that this limit could be overcome using optical near-fields. The key to utilizing optical near-fields is to realize novel nanometric fabrication and device operation by the control of an intrinsic interaction between nanometer-sized materials via optical near-fields. This has not been realized using conventional optical science and technology. This novel field of science and technology is nanophotonics.

One decade after the editor commenced his pioneering research into optical near-fields, a reliable technology was established for fabricating high-quality fiber probes. This led to the development of near-field optical microscopy and spectroscopy, with high resolution, beyond the diffraction limit of conventional optical microscopy. After establishing the fiber-probe technology, the nature of optical near-fields was studied by regarding the optical near-field as an electromagnetic field that mediates the interaction between nanometric materials. As a result, the physically intuitive concept of a *dressed photon* was established to describe optical near-fields, i.e. the interaction between nanometric materials is mediated by exchanging dressed photons.

After starting to develop a novel theory to describe this interaction, it was found that optical near-fields, i.e. dressed photons, could be used to realize novel photonic devices, fabrication techniques, and systems. Therefore, in 1993, the idea of nanophotonics was proposed. This is a novel technology that utilizes the optical near-field to realize novel devices, fabrications, and systems. Following elaboration of the idea of nanophotonics, much theoretical and experimental work has been carried out, and several novel functions and phenomena have been discovered in device operation and fabrication techniques. The objective of this book is to review the innovations of nanofabrication using nanophotonics.

In conventional optical science and technology, light and matter are discussed separately, and the flow of optical energy in a photonic system is considered unidirectional, from a light source to a photodetector. By contrast, in nanophotonics, light and matter have to be regarded as being coupled to each other and the energy flow between nanometric particles is bidirectional. This means that nanophotonics should be regarded as a *technology fusing optical fields and matter*. The term nanophotonics is occasionally used for photonic crystals, plasmonics, metamaterials, silicon photonics, and quantum-dot lasers using conventional propagating light, although they are not based on optical near-field interactions. The development of nanophotonics requires far-reaching physical insights into the local electromagnetic interaction in the nanometric subsystem composed of electrons and photons.

Chapter 1 of this book reviews the history, background, and present status of research and development in nanophotonics, including its application to nanofabrication; it also comments on future perspectives. Chapter 2 presents the principles of nanofabrication based on dressed-photon models, describes adiabatic and non-adiabatic processes in nanofabrication, and demonstrates their application to chemical vapor deposition and lithography.

Chapters 3–12 review practices of nanofabrication: Chapter 3 deals with nanofabrication using self-organization and related technology in order to control the size and position of the fabricated nanometric materials. Chapter 4 describes a method of fabricating semiconductor quantum dots, which is an important fundamental technology for realizing nanophotonic devices. Chapter 5 describes the optical properties of a ZnO nanorod heterostructure, which is also a key material for nanophotonic device applications. Chapter 6 discusses lithography based on nanophotonics, which was developed by industry. The details of a lithography system are reviewed and various fabricated patterns are presented. Chapter 7 deals with the fabrication of FePt nanopatterned media for high-density magnetic storage. It also reviews the technology of block-copolymer lithography to be used for this fabrication. Chapter 8 reviews a nanophotonics recording device for a high-density storage system that increases the storage data density to 1 Tbit/inch^2. Chapter 9 reviews the performances of X-ray devices that were fabricated using nanophotonic lithography; their high-quality performance is demonstrated and, thus, the potential of nanophotonic lithography as a novel nanofabrication tool may be recognized. Chapter 10 is devoted to describing periodic nanostructure formation on hard thin films using femtosecond laser ablation. Chapter 11 reviews a novel nanophotonic waveguide composed of quantum dots with low-loss optical energy transmission and low crosstalk. It describes the principle of operation, modeling, fabrication process, and performance of the waveguide. The final chapter concerns the intrinsic characteristics of hierarchy in optical near-fields and its application to nanofabrication, such as generating smaller-scale structures from larger-scale ones via optical near-field interactions. Each chapter is written by leading scientists in the relevant field. Consequently, I hope that high-quality scientific and technical information is provided to scientists, engineers, and students who are, and will be engaged in, nanofabrication and its applications.

Tokyo, September 2008 *Motoichi Ohtsu*

List of Contributors

Fumiko Akagi
Central Research Laboratory
Hitachi, Ltd.
Japan

Kouichi Akahane
National Institute of Information and
Communications Technology
4-2-1 Nukui-Kita
Koganei
Tokyo 184-8795
Japan

Hiroyuki Hieda
Research and Development Center
Toshiba Corp.
Japan

Kazuya Hirata
SIGMA KOKI Co., Ltd.
Japan

Hirokazu Hori
University of Yamanashi
4-3-11 Takeda
Kofu
Yamanashi 400-8511
Japan

Takashi Imazono
Japan Atomic Energy Agency
8-1 Umemidai Kizugawa 619-0215
Japan

Yasuhisa Inao
Canon Research Center
Canon Inc.
30-2 Shimomaruko 3-chome
Ohta-ku
Tokyo 146-8501
Japan

Toshiki Ito
Canon Research Center
Canon Inc.
30-2 Shimomaruko 3-chome
Ohta-ku
Tokyo 146-8501
Japan

Tadashi Kawazoe
University of Tokyo
2-11-16 Yayoi
Bunkyo-ku
Tokyo 113-8656
Japan

Kokoro Kitamura
The University of Tokyo
2-11-16 Yayoi
Bunkyo-ku
Tokyo 113-8656
Japan

Nanophotonics and Nanofabrication. Edited by Motoichi Ohtsu
Copyright © 2009 WILEY-VCH Verlag GmbH & Co. KGaA, Weinheim
ISBN: 978-3-527-32121-6

Masato Koike
Japan Atomic Energy Agency
8-1 Umemidai
Kizugawa 619-0215
Japan

Ryo Kuroda
Canon Research Center
Canon Inc.
30-2 Shimomaruko 3-chome
Ohta-ku
Tokyo 146-8501
Japan

Lih Y. Lin
Electrical Engineering Department
University of Washington
Seattle
WA 98195-2500
USA

Takuya Matsumoto
Central Research Laboratory
Hitachi, Ltd.
Japan

Shinji Miyauchi
Shimadzu Corp.
1 Nishinokyo-Kuwabaracho
Nakagyo-ku
Kyoto 604-8511
Japan

Kenzo Miyazaki
Advanced Laser Research Section
Institute of Advanced Energy
Kyoto University
Gokasho
Uji
Kyoto 611-0011
Japan

Natsuhiko Mizutani
Canon Research Center
Canon Inc.
30-2 Shimomaruko 3-chome
Ohta-ku
Tokyo 146-8501
Japan

Shinji Nakazato
Canon Research Center
Canon Inc.
30-2 Shimomaruko 3-chome
Ohta-ku
Tokyo 146-8501
Japan

Makoto Naruse
National Institute of Information and
Communications Technology
4-2-1 Nukui-kita
Koganei
Tokyo 184-8795
Japan

and

The University of Tokyo
2-11-16 Yayoi
Bunkyo-ku
Tokyo 113-8656
Japan

Tetsuya Nishida
Central Research Laboratory
Hitachi, Ltd.
Japan

Wataru Nomura
School of Engineering
The University of Tokyo
Japan

Motoichi Ohtsu
University of Tokyo
Dept. of Electronics Engineering
2-11-16 Yayoi
Bunkyo-ku
Tokyo 113-8656
Japan

Kazuo Sano
Japan Atomic Energy Agency
8-1 Umenidai
Kizugawa 619-0215
Japan

Yoshinori Tabata
SIGMA KOKI Co., Ltd.
Japan

Akira Terao
Canon Research Center
Canon Inc.
30-2 Shimomaruko 3-chome
Ohta-ku
Tokyo 146-8501
Japan

Chia-Jean Wang
Electrical Engineering Department
University of Washington
Seattle
WA 98195-2500
USA
*Currently with Intellectual Ventures, Inc.

Tomohiro Yamada
Canon Research Center
Canon Inc.
30-2 Shimomaruko 3-chome
Ohta-ku
Tokyo 146-8501
Japan

Takako Yamaguchi
Canon Research Center
Canon Inc.
30-2 Shimomaruko 3-chome
Ohta-ku
Tokyo 146-8501
Japan

Naokatsu Yamamoto
National Institute of Information
and Communications Technology
4-2-1 Nukui-Kita
Koganei
Tokyo 184-8795
Japan

Takashi Yatsui
School of Engineering
The University of Tokyo
2-11-16 Yayoi
Bunkyo-ku
Tokyo 113-8656
Japan

Gyu-Chul Yi
National Creative Research Initiative
Center for Semiconductor Nanorods
and Department of Materials Science
and Engineering
POSTECH
Pohang 790-784
Korea

Present address:

Department of Physics and Astronomy
Seoul National University
Seoul 151-742
Korea

1
Introduction

Motoichi Ohtsu

1.1
History

Nanophotonics is a novel technology that utilizes the optical near-field, which is the electromagnetic field that mediates the interaction between nanometric particles located in close proximity to each other. The true nature of nanophotonics is to realize 'qualitative innovation' in photonic devices, fabrication techniques, and systems by utilizing novel functions and phenomena caused by optical near-field interactions, which are impossible as long as conventional propagating light is used. The author first proposed nanophotonics in 1993 as a way to transcend the diffraction limit, which impedes reducing the size of photonic devices, to improve the resolution of optical fabrication techniques, and increasing the storage density of optical disk memories [1]. Based on his proposal, the Optical Industry Technology Development Association (OITDA) of Japan organized the nanophotonics technical group, and intensive discussions on the future direction of nanophotonics started in April 1994, in collaboration with academia and industry. Although photonic crystals, plasmonics, metamaterials, silicon photonics, and quantum dot lasers have been popular subjects of study in recent years, they are all based on diffraction-limited wave optics. Even if novel or nanometer-sized materials are used for these subjects, the size of a photonic device cannot be reduced beyond the diffraction limit as long as propagating light is used for its operation.

This chapter describes the history and present activities of nanophotonics. Before nanophotonics was founded, the study of optical near-fields started in Japan in the early 1980s, separate from European and American research [2]. Few attended when the author was invited to the first international workshop on near-field optics in 1992 [3], but the numbers have increased very rapidly in subsequent conferences, and near-field optics has become very popular worldwide.

Although the number of near-field optics researchers increased dramatically in the 1990s, almost all of them focused on microscopy or spectroscopy. Unlike them, intensive research and development of novel devices, fabrication techniques, and systems was conducted after nanophotonics was proposed in 1993. To support these studies, a novel theoretical model of optical near-fields was established and applied to the design of nanophotonic devices, fabrication techniques, and systems. For these designs, local energy transfer and its subsequent dissipation are indispensable. They are possible only using optical near-fields [4], which are the elementary surface excitations on nanometric particles, or in the other words, *dressed photons* that carry the material energy. Nanophotonics evolved into related sciences, one of which is atom photonics, which studies the control of the thermal motions of neutral atoms in a vacuum using optical near-fields [5]. The above-mentioned theoretical model has been used as a design criterion for atom photonics for opening technologies in atomic-level material fabrication. Furthermore, atom photonics has triggered related research, such as that seen in the atom chip [6].

1.2
Fiber Probes and Sensing Systems

One decade after near-field optics started in the early 1980s, a reliable, reproducible, and selective chemical etching technology was established for fabricating high-quality fiber probes [7]. This led to the realization of a high-resolution probe with a 1-nm apex radius, a high-efficiency probe with 10% optical near-field generation efficiency, and other devices. Elsewhere in the world, methods that heat and pull the fiber were most popular at that time [8], although the performance of chemically etched fiber probes was superior. The methodology used to fabricate these probes was transferred to industry, and Japanese companies started producing high-quality commercial fiber probes. A near-field optical microscope was developed using these fiber probes, and a variety of ultrahigh-resolution images have been achieved, such as one of a single-stranded DNA molecule with a resolution greater than 4 nm [9], which is a world record.

A near-field spectrometer was also developed for diagnosing single semiconductor quantum dots [10], semiconductor devices [11], a single organic molecule [12], and biological specimens [13]. Many experimental results on spatially resolved photoluminescence and Raman spectra with a 10-nm resolution have been accumulated [14]. Patents have been transferred to industry and a Japanese company has produced commercial near-field photoluminescence spectrometers operating in the ultraviolet–infrared and liquid helium–room temperature ranges [15]. They are popular in a variety of nanoscience and technology fields. Instead of using fiber probes, an apertureless probe was sometimes used because of its simplicity of fabrication and the possibility of field enhancement. However, it was demonstrated that apertureless probes do not realize high resolution due to the scattering of residual propagating light [16].

1.3 Theory

The use of optical near-fields was proposed about 80 years ago as a way to break the diffraction limit [17]. This proposal holds that an optical near-field can be generated on a subwavelength-sized aperture by irradiating the propagating light. It also holds that the size of the spatial distribution of the optical near-field energy depends not on the wavelength of the incident light, but on the aperture size. However, in the early stage of such studies, the concept of optical near-fields was not clearly differentiated from that of an evanescent wave on a planar material surface (i.e. a two-dimensional topographical material) or that of a guided wave in a subwavelength-sized cross-sectional waveguide (i.e. a one-dimensional topographical material). To distinguish these clearly, note that an evanescent wave is generated by primary excitations, that is, electronic dipoles induced near the two-dimensional material surface, which align periodically depending on the spatial phase of the incident light. In contrast, the guided wave in a subwavelength-sized cross-sectional waveguide is generated by the electronic dipoles induced along the one-dimensional waveguide material. They align periodically depending on the spatial phase of the incident light. Silicon waveguides used for silicon photonics and metallic waveguides used for plasmonics are examples. The two-dimensional evanescent wave and one-dimensional guided wave are both diffraction-limited light waves because they are generated by the periodic alignment of electric dipoles depending on the spatial phase of the incident light.

Unlike these waves, an optical near-field is generated by electronic dipoles induced in a nanometric particle (i.e. a subwavelength-sized zero-dimensional topographical material). Their alignment is independent of the spatial phase of the incident light because the particles are much smaller than the wavelength of the incident light. Instead, the optical near-field depends on the size, conformation, and structure of the particles. Due to this independence and dependence, optical science and technology beyond the diffraction limit can be realized only by an optical near-field, and not by an evanescent wave or a guided wave.

Methods such as Green's function, a calculation using the finite-difference time-domain method, etc. have been developed to describe the optical near-field semi-quantitatively based on conventional optics theories [18]. However, conventional optics theories do not provide any physically intuitive pictures of *nonpropagating* nanometric optical near-fields because these theories were developed to describe only the light waves *propagating* through macroscopic space or materials. For nanophotonics, a novel theory has been developed based on a framework that is completely different from those of the conventional theories. This novel theory is based on the interaction and energy transfer between nanometric particles via an optical near-field. This perspective is essential because the interaction, energy transfer, and subsequent dissipation are indispensable for nanophotonic devices and fabrications. That is, to observe a nonpropagating optical near-field, a second particle is inserted to generate observable scattered light by disturbing the optical near-field. However, the real system is more complicated because the *nanometric subsystem* (the two particles and

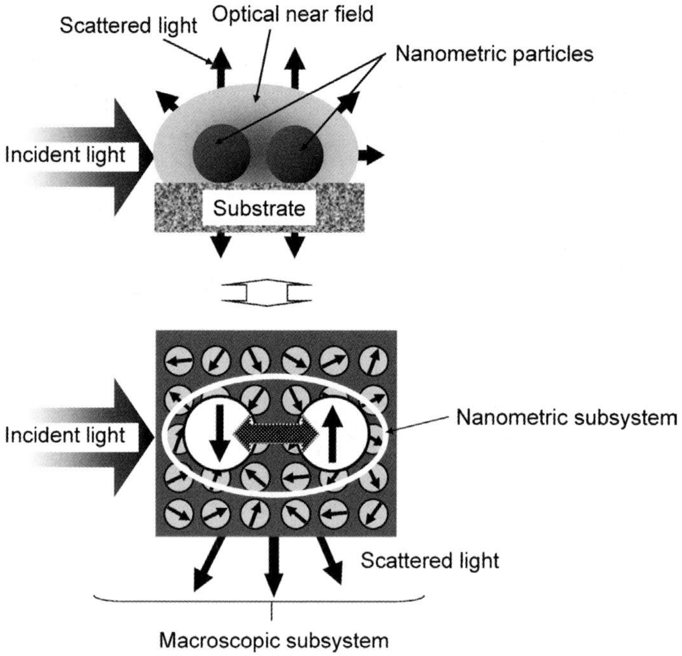

Figure 1.1 A nanometric subsystem composed of two particles and an optical near-field, which are buried in a macroscopic subsystem.

the optical near-field) is buried in a *macroscopic subsystem* consisting of the macroscopic substrate material and the macroscopic electromagnetic fields of the incident and scattered lights (see Figure 1.1).

The premise behind the novel theory is to avoid the complexity in describing all of the behaviors of nanometric and macroscopic subsystems rigorously, since we are interested only in the behavior of the nanometric subsystem. The macroscopic subsystem is expressed as an exciton–polariton, which is a mixed state of material excitation and electromagnetic fields. Since the nanometric subsystem is excited by an electromagnetic interaction with the macroscopic subsystem, the projection operator method is effective for describing the quantum-mechanical states of these systems [4, 19]. Under this treatment, the nanometric subsystem is regarded as being isolated from the macroscopic subsystem, while the functional form and magnitude of effective interactions between the elements of the nanometric subsystem are influenced by the macroscopic subsystem. That is, the two nanometric particles can be considered as being isolated from the surrounding macroscopic system and as interacting by exchanging exciton–polariton energies. Therefore, the optical near-fields can be considered as *dressed photons*, which carry the material energy.

This local electromagnetic interaction takes place so quickly that the uncertainty principle can allow the exchange of a virtual exciton–polariton nonresonantly, as well as the exchange of a real exciton–polariton resonantly (see Figure 1.2).

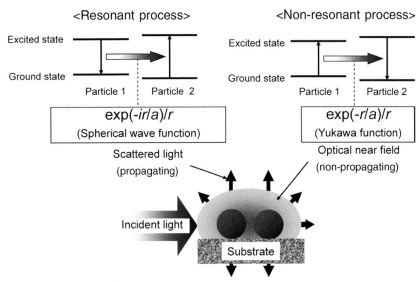

Figure 1.2 Resonant and nonresonant processes.

The interaction in the nonresonant process, corresponding to mediation by the optical near-field, is expressed by a Yukawa function that represents the localization of the optical near-field around the nanometric particles like an electron cloud around an atomic nucleus. Its decay length is equivalent to the material size [4]. On the other hand, the interaction corresponding to the resonant process is expressed by a conventional spherical wave function, and indicates the mediation by a conventional propagating field.

As described above, the optical near-field is an electromagnetic field that mediates the interaction between nanometric particles located in close proximity to each other. Nanophotonics is the technology utilizing this field to realize novel devices, fabrications, and systems. That is, a photonic device with a novel function can be operated by transferring the optical near-field energy between nanometric particles and its subsequent dissipation. In such a device, the optical near-field transfers a signal and carries the information. Novel photonic systems become possible by using these novel photonic devices. Furthermore, if the magnitude of the transferred optical near-field energy is sufficiently large, the structures or conformations of nanometric particles can be modified, which suggests the feasibility of a novel photonic fabrication technique.

With these treatments, a physically intuitive picture of nonpropagating nanometric optical near-fields was established. On reading this, one may understand that the true nature of nanophotonics is to realize 'qualitative innovation' in photonic devices, fabrications, and systems by utilizing novel functions and phenomena caused by optical near-field interactions, which are impossible as long as conventional propagating light is used. Nanophotonics undoubtedly has the advantage of exceeding the diffraction limit of light, that is, 'quantitative innovation'.

This innovation is a really challenging issue in quantum theory because it is the first time the structure, interaction, the temporal evolution, etc. in the region much less than the wavelength, or equivalently those in the region much less than the de Broglie wavelength, has been discussed. This is the first encounter for people to work in such conditions. However, it should be pointed out that this innovation is only a secondary feature of nanophotonics, as compared with the 'qualitative innovation'. Qualitative innovation has been already realized for devices, fabrications, and systems by appropriately utilizing the true nature of the optical near-field interaction, and some of these are described in the following sections.

1.4
Devices

For nanophotonic device operation, it is essential to transfer the signal and to fix the transferred signal magnitude at the output terminal, which can be achieved by transferring the optical near-field energy and its subsequent dissipation, respectively [20]. Several research groups have recently begun similar discussions on this operation (e.g., refer to [21]). The exciton dynamics of a three-quantum-dot system, as well as a system consisting of a pair of quantum dots, have also been discussed for model nanophotonic functional devices, based on the above formulation and a quantum master equation. The recent analysis includes spin polarization and excitation transfer [22].

As an example, a pair of closely spaced equivalent quantum dots is used as the input terminal of the nanophotonic device (see Figure 1.3). As a result of the optical near-field interaction between the two quantum dots driven by the input optical signal, the quantized energy levels of the exciton in the quantum dots are split in two. One half corresponds to the symmetric state of the exciton, and the other is the antisymmetric state. They represent the respective parallel and antiparallel electric dipole moments induced in the two quantum dots. A third larger quantum dot, located near to the input terminal (see Figure 1.4), is used as the output terminal of the device. The higher energy level of the exciton in this quantum dot is tuned to that of the symmetric state of the input terminal, which is possible by adjusting the size of the third quantum dot. As a result of this tuning, optical near-field energy

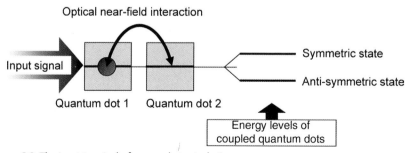

Figure 1.3 The input terminal of a nanophotonic device.

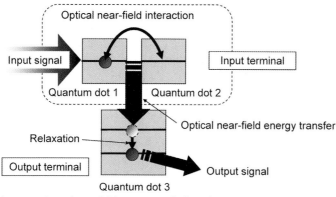

Figure 1.4 Optical near-field energy transfer from the input to the output terminal and subsequent dissipation.

can be transferred from the input to the output terminal, allowing signal transfer. The excitation transferred to the output terminal is dissipated in the third quantum dot immediately by coupling with phonons, which fixes the magnitude of the transferred signal.

A device utilizing the symmetric state of the input terminal in the manner shown in Figure 1.4 is called a 'phonon-coupled device.' Examples of phonon-coupled devices include nanophotonic switches [23], logic gates, such as AND and NOT gates [24], content-addressable memory [25], and digital-to-analog converters [26] (see Figures 1.5(a)–(c)). Conversely, a device utilizing the antisymmetric state is called a 'propagating light-coupled device,' and examples include an optical buffer memory and super-radiant-type optical pulse generators [27]. They result in qualitative innovative device operations that are impossible when conventional propagating light is used. The nanophotonic switch can be as small as 15 nm. The dynamic behavior of its output signal agrees well with the calculated results based on the dressed-photon model. The figure of merit for this device is 10–100 times larger than those of conventional photonic switches operated using propagating light. The magnitude of the heat dissipated from the nanophotonic switch is estimated to be as low as 10^{-12} W in the case of repetitive 1-GHz operation, which is only 10^{-5} times that of a conventional semiconductor transistor. For the lowest heat dissipation, a device operated using a single photon has also been recently demonstrated [28]. Such a large figure of merit and the ultralow heat dissipation suggest a variety of applications to novel optical computation and information processing systems.

To connect nanophotonic and macroscopic photonic devices, a novel device has been developed that concentrates the propagating light energy to a nanometric region [29] (see Figure 1.6). This device has been named an 'optical nanofountain' and its equivalent numerical aperture is as high as 40. Although several plasmonic devices have been developed for this interconnection [30, 31], they cannot realize such a large numerical aperture because the plasmonics method utilizes the classical wave-optical picture. The letters 'on' in the word 'plasmon' represent the quanta,

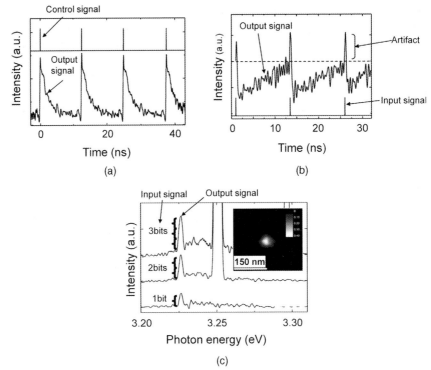

Figure 1.5 The operation of nanophotonic devices. (a) Temporal behavior of the output signal from a nanophotonic switch [23]. (b) Temporal behavior of the output signal from a NOT gate [24]. (c) Spectral intensity of the output signal from a three-bit digital-to-analog converter [26].

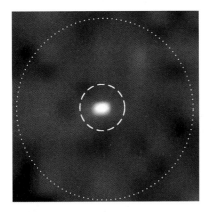

Figure 1.6 Spatial profile of the output signal intensity from an optical nanofountain [29]. The outer circle represents the spot of the incident propagating light.

i.e. the quantum-mechanical picture of the plasma oscillation of free electrons in a metal. Instead of this picture, plasmonics utilizes the classical wave-optical picture by using conventional terminology, such as the refractive index, wave number, and guided mode. Even when a metal is irradiated with light that obeys quantum mechanics, the quantum-mechanical property is lost because the light is converted into the plasma oscillation of electrons, which has a short-phase relaxation time. Therefore, as long as plasma oscillation is used, it is impossible to reduce the device size over the diffraction limit.

1.5 Fabrications

Novel optical nanofabrication techniques beyond the diffraction limit are required for producing a variety of conventional electronic/photonic devices and nanophotonic devices. To fabricate nanophotonic devices, several capabilities are required. For example, a variety of materials must be deposited on a substrate, and the inaccuracy of their sizes and positions must be as low as 1 nm for efficient reproducible optical near-field energy transfer. However, conventional fabrication technologies using electron beams, ion beams, and propagating light cannot meet these requirements due to their low resolution, contamination of and damage to the substrate, and low throughput. To meet the requirements, novel technologies had to be developed, and these have been realized by utilizing optical near-field energy transfer. As representative examples of such nanophotonic fabrication techniques, photochemical vapor deposition and photolithography are described in this section.

Photochemical vapor deposition is a way to deposit materials on a substrate using a photochemical reaction with ultraviolet light that predissociates metalorganic molecules by irradiating gaseous molecules or molecules adsorbed on the substrate. Consequently, the electrons in the molecules are excited to a higher energy level following the Franck–Condon principle. This is an adiabatic process because no molecular vibrations or rotations are excited; that is, the Born–Oppenheimer approximation is valid. After excitation, the electron transits to the dissociative energy level. As a result of this transition, the molecule is dissociated and the atoms forming the molecule are deposited on the substrate. By using an optical near-field as the light source for this deposition scheme, nanometric materials can be fabricated whose size and position are controlled accurately by the spatial distribution of the optical near-field energy. Nanometric metallic particles of Zn and Al and light-emitting semiconductor particles (e.g., ZnO and GaN) have been deposited with size and positional accuracies far beyond the diffraction limit [32–36]. This is an example of a quantitative innovation.

Although ultraviolet light with high photon energy must be used for the abovementioned adiabatic process, it has been found that an optical near-field with a much lower photon energy (i.e. visible light) can dissociate the molecule. This has been explained by a theoretical model of the virtual exciton–polariton exchange between a metalorganic molecule and the fiber probe tip used to generate the optical near-field. In other words, this exchange excites not only the electron, but also molecular vibration. This is a nonadiabatic process, which does not follow the Franck–Condon principle, and therefore the Born–Oppenheimer approximation is no longer valid. Several experimental results have been reported, including the nonadiabatic dissociation of diethylzinc (DEZn) molecules by a visible optical near-field and the deposition of nanometric Zn particles [37]. The virtual exciton–polariton model explains this nonadiabatic process quantitatively by introducing the contribution of a phonon excited in the fiber probe tip [38]. A theoretical model for a pseudo-one-dimensional optical near-field probe system has led to a simple understanding of this process [39].

The nonadiabatic process presented above suggests that large, expensive ultraviolet light sources are no longer required, although they have long been indispensable for conventional photochemical vapor deposition. It also suggests that the process can dissociate optically inactive molecules (i.e. inactive to the propagating light), which is advantageous for protecting the environment because most optically inactive molecules are stable chemically and harmless. For example, optically inactive $Zn(acac)_2$ molecules have been dissociated nonadiabatically using an optical near-field to deposit nanometric Zn particles [40].

Photolithography is a technology used to carve a substrate material. After coating the substrate with a thin film of a photoresist, light is irradiated through a photomask to induce a photochemical reaction in the photoresist. When the aperture on the photomask is smaller than the wavelength of the light, the transmission of the propagating light is sufficiently low, while an optical near-field is generated at the aperture. Using the photochemical reaction between the optical near-field and photoresist, a nanometric pattern beyond the diffraction limit is formed on the photoresist, and a chemical etching process is subsequently used to carve the substrate. Several preliminary experiments have used a fiber probe to generate optical near-fields [41]. Alternatively, a photomask is used to improve the throughput of fabrication dramatically. Practical technologies [42], such as using a two-layered photoresist, have been developed to form deep patterns, thus realizing a quantitative innovation [43, 44].

As long as an adiabatic process is used for photolithography, an ultraviolet light with a high photon energy is required to induce the photochemical reaction via the optical near-field. However, photolithography using a nonadiabatic process is possible, as in the above-mentioned case of photochemical vapor deposition; that is, a photochemical reaction can even be induced using visible light with a very low photon energy. Narrow corrugated patterns were fabricated using nonadiabatic photolithography [45]. This has been analyzed theoretically based on a virtual exciton–polariton model by introducing the contribution of phonons. This result represents a quantitative innovation in photolithography, suggesting that large, expensive ultraviolet light sources are no longer required, and that harmless, chemically stable molecules can be used as the photoresist, even if they are optically inactive. An example of an optically inactive resist film is the one used for electron-beam lithography. A photochemical reaction is induced in this film via a nonadiabatic process, and fine patterns have been fabricated using a photomask consisting of a two-dimensional array of circular disks [45].

Fabrication using adiabatic processes suffers from the contributions of the low-intensity propagating light transmitted through the aperture on the photomask and of the plasmon generated on the photomask, which limit the resolution of the fabrication. In contrast, since the nonadiabatic process is free of these contributions, greater resolution, higher contrast, and a variety of patterns (e.g., Ts, Ls, and rings) have been realized [45]. A commercial prototype of a compact desktop machine that can fabricate corrugations with a 20–50-nm line width over a 25×25-mm substrate area has been developed [44].

1.6
Applications to Systems and Evolution to Related Sciences

Nanophotonics has realized quantitative and qualitative innovations. As examples of its evolution to related science and technology, this section describes applications to novel photonic systems and atom photonics. Nanophotonic systems utilize nanophotonic devices. For example, novel architectures have been proposed for optical signal-transmission systems and their performance has been confirmed experimentally. They include interconnections and summation architecture involving an optical nanofountain [25], computing using nanophotonic switches and an optical nanofountain [26], and data broadcasting using multiple nanophotonic switches [46] (see Figure 1.7). They have realized quantitative innovations by decreasing the device size and power consumption beyond the diffraction limit. More importantly, qualitative innovation has been realized because of the novel functions of nanophotonic devices, which are impossible using conventional photonic devices. Quantitative innovation has already been realized by breaking the diffraction limit of optical/magnetic hybrid disk storage density [47].

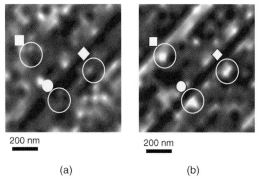

(a)   (b)

Figure 1.7 The output signal intensity from a data broadcast using three nanophotonic switches [46]. (a) The switches are off. (b) The switches are on.

Qualitative innovation has also been proposed by applying the hierarchy in optical near-fields to memory retrieval [48].

An example of evolution to a related science is atom photonics, which controls the thermal motions of neutral atoms in a vacuum using optical near-fields [5]. Theoretical studies have examined single-atom manipulation based on the virtual exciton–polariton model [19], and experimental studies have involved the first successful guidance of an atom through a hollow optical fiber [49]. Recent studies have examined atom-detecting devices [50], atom deflectors [51], and an atomic funnel [52]. Atom photonics will open up a new field of science that examines the interaction between dressed photons and a single atom.

Basic research for progress in nanophotonics is being actively carried out. An optical near-field problem has been formulated in terms of the Carniglia–Mandel model as a complete and orthogonal set that satisfies the infinite planar boundary conditions between the dielectric and a vacuum. This approach has revealed interesting atomic phenomena occurring near the surface, which have been analyzed based on angular spectrum representation [53, 54]. For example, optical radiation from an excited molecule on the substrate surface has been analyzed, which agreed quantitatively with experimental results [55]. A self-consistent, nonlocal, semiclassical theory on light–matter interactions has been developed to discuss the optical response of a variety of nanostructures [56]. In particular, the size dependence and allowance of a dipole-forbidden transition in a nanometric quantum-dot system were noted [57, 58]. The optical manipulation of nanometric objects in superfluid ^{4}He has been investigated based on the nonlocal semiclassical theory [59]. Electron transport through molecular bridges connecting nanoscale electrodes has been formulated [60], and a unified method has been proposed to treat extended and polaron-like localized states coupled with molecular vibrations. A one-dimensional molecular bridge made of thiophene molecules has been analyzed numerically. The study of optical near-fields associated with molecular bridges is now in progress. In addition, as basic experimental work, desorption and ionization have been carried out assisted by optical near-fields, and their application to mass spectroscopy has been proposed [61, 62].

1.7
Toward the Future

This chapter reviewed the history and recent progress in nanophotonics, a novel optical technology proposed by the author. It utilizes the local interaction between nanometric particles via optical near-fields. The optical near-fields are the elementary surface excitations on nanometric particles, that is *dressed photons* that carry the material energy. This chapter emphasized that the true nature of nanophotonics is to realize 'qualitative innovation' in photonic devices, fabrication techniques, and systems by utilizing novel functions and phenomena caused by optical near-field interactions, which are impossible as long as conventional propagating light is used. Evidence of such innovation was described, that is, novel devices such as switches, logic gates, and a nanofountain utilizing optical near-fields as a carrier to transmit the signal. Novel fabrications that utilized a nonadiabatic photochemical reaction with visible (not ultraviolet) optical near-fields were also demonstrated, as well as novel nanophotonic information and communication systems that combined nanophotonic devices. Furthermore, it was noted that nanophotonics has evolved into atom photonics.

Nanophotonics is now a key optical technology. However, the name 'nanophotonics' is occasionally used for photonic crystals, plasmonics, metamaterials, silicon photonics, and quantum-dot lasers using conventional propagating light. Here, one should consider a stern warning by C. Shannon on the casual use of the term 'information theory,' which was a trend in the study of information theory during the 1950s [63]. The term 'nanophotonics' has been used in a similar way, although some work in 'nanophotonics' is not based on optical near-field interactions. For the true development of nanophotonics, one needs deep physical insight into the *dressed photons* and the nanometric subsystem composed of electrons and photons.

References

1 Ohtsu, M. (1993) *Progress in Nano-Electro-Optics V*, Springer-Verlag, Berlin. pp. VII–VIII (Based on 'nanophotonics' proposed by Ohtsu in 1993 OITDA Optical Industry Technology Development Association Japan organized the nanophotonics technical group in 1994 and discussions on the future direction of nanophotonics were started in collaboration with academia and industry).

2 Ohtsu, M. (2000) *Near-Field Optics: Principles and Applications*, World Scientific Publishing Co, Singapore. pp. 1–8.

3 Pohl D.W. and Courjon D. (eds.), (1993) *Near-field Optics*, Kluwer Academic Publishers, Dordrecht. pp. 1–410.

4 Ohtsu, M. and Kobayashi, K. (2004) *Optical Near-fields*, Springer-Verlag, Berlin. pp. 109–120.

5 Ito, H. (1998) *Near-Field Nano/Atom Optics and Technology*, Springer-Verlag, Berlin. pp. 217–293.

6 Brugger, K., Calarco, T., Cassettari, D., Folman, R., Haase, A., Hessmo, B., Kruger, P., Maier, T. and Schmiedmayer, J. (2000) Nanofabricated atom optics: Atom chips. *Journal of Modern Optics*, **47** (14), 2789–2809.

7 Mononobe, S. (1998) *Near-Field Nano/Atom Optics and Technology*, Springer-Verlag, Berlin. pp. 31–69.

8 Betzig, E. and Trautman, J.K. (1992) Near-field optics: Microscopy, spectroscopy, and surface modification beyond the diffraction limit. *Science*, **257** (5067), 189–195.

9 Maheswari Rajagopalan, U., Mononobe, S., Yoshida, K., Yoshimoto, M. and Ohtsu, M. (1999) Nanometer level resolving near-field optical microscope under optical feedback in the observation of a single-string deoxyribo Nucleic Acid. *Japanese Journal of Applied Physics*, **38** (12A), 6713–6720.

10 Matsuda, K., Saiki, T., Saito, H. and Nishi, K. (2000) Room temperature photoluminescence spectroscopy of a single $In_{0.5}Ga_{0.5}As$ quantum dot by using highly sensitive near-field scanning optical microscope. *Applied Physics Letters*, **76** (1), 73–75.

11 Fukuda, H., Saiki, T. and Ohtsu, M. (2001) Diagnostics of semiconductor devices beyond the diffraction limit of light. *Sensors and Materials*, **13** (8), 445–460.

12 Hosaka, N. and Saiki, T. (2001) Near-field fluorescence imaging of single

molecules with a resolution in the range of 10 nm. *Journal of Microscopy*, **202** (2), 362–364.
13. Ushiki, T. (2005) Scanning Near-Field Optical/Atomic Force Microscopy in Biology. presented at the 5th Asia-Pacific Conference on Near-Field Optics, 15–17 November, Niigata, Japan, Paper 2–1, Optical Society of Japan.
14. Narita, Y., Tadokoro, T., Ikeda, T., Saiki, T., Mononobe, S. and Ohtsu, M. (1998) Near-field Raman spectral measurement of polydiacetylene. *Applied Spectroscopy*, **52** (9), 1141–1144.
15. Sato, F. (2004) Development of time resolved near-field optical spectrometer by time correlated single photon counting (TCSPC-NSOM). *Jasco Report*, **46** (2), 34–38, (in Japanese).
16. Hubert, C., Lerondel, G., Bachelot, R., Kostcheev, S., Grand, J., Vial, A., Barciesi, D., Royer, P., Chang, S.H., Gray, S.K., Wiederrecht, G.P. and Shatz, G.C. (2005) Near-Field Photochemical Imaging of Noble Metal Nano-Object Using an Azodyne Polymer: From Single to Near-Field Coupled Objects. presented at the 5th Asia–Pacific Conference on Near-Field Optics, 15–17 November, Niigata, Japan, Paper 3–4, Optical Society of Japan.
17. Synge, E.H. (1928) A suggested method for extending microscopic resolution into the ultra-microscopic region. *Philosophical Magazine and Journal of Science*, **6** (7), 356–362.
18. Ohtsu, M. and Kobayashi, K. (2004) *Optical Near-fields*, Springer-Verlag, Berlin, pp. 88–108.
19. Kobayashi, K., Sangu, S., Ito, H. and Ohtsu, M. (2001) Near-field optical potential for a neutral atom. *Physical Review A*, **63** (1), 013806.
20. Sangu, S., Kobayashi, K., Shojiguchi, A. and Ohtsu, M. (2004) Logic and functional operations using a near-field optically coupled quantum-dot system. *Physical Review B-Condensed Matter*, **69** (11), 115334.
21. Achermann, M., Petruska, M.A., Kos, S., Smith, D.L., Koleske, D.D. and Klimov, V.I. (2004) Energy-transfer pumping of semiconductor nanocrystals using an epitaxial quantum well. *Nature*, **429**, 642–646.
22. Sato, A., Minami, F. and Kobayashi, K. (2007) Spin and excitation energy transfer in a quantum-dot pair system through optical near-field interactions. *Physica*, **E40**, 313–316.
23. Kawazoe, T., Kobayashi, K., Sangu, S. and Ohtsu, M. (2003) Demonstration of a nanophotonic switching operation by optical near-field energy transfer. *Applied Physics Letters*, **82** (18), 2957–2959.
24. Kawazoe, T., Kobayashi, K., Akahane, K., Naruse, M., Yamamoto, N. and Ohtsu, M. (2006) Demonstration of nanophotonic NOT gate using near-field optically coupled quantum dots. *Applied Physics B-Lasers and Optics*, **84** (1–2), 243–236.
25. Naruse, M., Miyazaki, T., Kubota, F., Kawazoe, T., Kobayashi, K., Sangu, S. and Ohtsu, M. (2005) Nanometric summation architecture based on optical near-field interaction between quantum dots. *Optics Letters*, **30** (2), 201–203.
26. Naruse, M., Miyazaki, T., Kawazoe, T., Kobayashi, K., Sangu, S., Kubota, F. and Ohtsu, M. (2005) Nanophotonic computing based on optical near-field interactions between quantum dots. *IEICE Transactions on Electronics*, **E88-** (9), 1817–1823.
27. Shojiguchi, A., Kobayashi, K., Sangu, S., Kitahara, K. and Ohtsu, M. (2003) Superradiance and dipole ordering of an *N* two-level system interacting with optical near-fields. *Journal of the Physical Society of Japan*, **72** (11), 2984–3001.
28. Kawazoe, T., Tanaka, S. and Ohtsu, M. (2008) Single-photon emitter using excitation energy transfer between quantum dots. *Journal of Nanophotonics*, **2** (10), 029502.

29 Kawazoe, T., Kobayashi, K. and Ohtsu, M. (2005) Optical nanofountain: A biomimetic device that concentrates optical energy in a nanometric region. *Applied Physics Letters*, **86** (10), 103102.

30 Yatsui, T., Kourogi, M. and Ohtsu, M. (2001) Plasmon waveguide for optical far/near-field conversion. *Applied Physics Letters*, **79** (27), 4583–4585.

31 Takahara, J., Yamaguchi, S., Taki, H., Morimoto, A. and Kobayashi, T. (1997) Guiding of a one-dimensional optical beam with nanometer diameter. *Optics Letters*, **22** (7), 475–457.

32 Polonski, V., Yamamoto, Y., Kourogi, M., Fukuda, H. and Ohtsu, M. (1999) Nanometric patterning of zinc by optical near-field photochemical vapour deposition. *Journal of Microscopy*, **194** (2/3), 545–551.

33 Yamamoto, Y., Kourogi, M., Ohtsu, M., Polonski, V. and Lee, G.H. (2000) Fabrication of nanometric zinc pattern with photodissociated gas-phase diethylzinc by optical near-field. *Applied Physics Letters*, **76**, 2173–2175.

34 Yatsui, T., Kawazoe, T., Ueda, M., Yamamoto, Y., Kourogi, M. and Ohtsu, M. (2002) Fabrication of nanometric single zinc and zinc oxide dots by the selective photodissociation of adsorption-phase diethylzinc using a nonresonant optical field. *Applied Physics Letters*, **81** (19), 3651–3653.

35 Lim, J., Yatsui, T. and Ohtsu, M. (2005) Observation of size-dependent resonance of near-field coupling between a deposited Zn dot and the probe apex during near-field optical chemical vapor deposition. *IEICE Transactions on Electronics*, **E88-C** (9), 1832–1835.

36 Yamamoto, Y., Kourogi, M., Ohtsu, M., Lee, G.H. and Kawazoe, T. (2002) Lateral integration of Zn and Al dots with nanometer-scale precision by near-field optical chemical vapor deposition using a sharpened optical fiber probe. *IEICE Transactions on Electronics*, **E85-C** (12), 2081–2085.

37 Kawazoe, T., Yamamoto, Y. and Ohtsu, M. (2001) Fabrication of a nanometric Zn dot by nonresonant near-field optical chemical-vapor deposition. *Applied Physics Letters*, **79** (8), 1184–1186.

38 Kawazoe, T., Kobayashi, K., Takubo, S. and Ohtsu, M. (2005) Nonadiabatic photodissociation process using an optical near-field. *Journal of Chemical Physics*, **122** (2), 024715.

39 Tanaka, Y. and Kobayashi, K. (2007) Optical near-field dressed by localized and coherent phonons. *Journal of Microscopy*, **229** (2), 228–232.

40 Kawazoe, T., Kobayashi, K. and Ohtsu, M. (2006) Near-field optical chemical vapor deposition using $Zn(acac)_2$ with a non-adiabatic photochemical process. *Applied Physics B-Lasers and Optics*, **84** (1–2), 247–251.

41 Smolyaninov, I.I., Mazzoni, D. and Davis, C.C. (1995) Near-field direct-write ultraviolet lithography and shear force microscopic studies of the lithographic process. *Applied Physics Letters*, **67** (26), 3859–3861.

42 Alkaisi, M.M., Blaikie, R.J., McNab, S.J., Cheung, R. and Cumming, D.R.S. (1999) Sub-diffraction-limited patterning using evanescent near-field optical lithography. *Applied Physics Letters*, **75** (22), 3560–3562.

43 Naya, M., Tsurusawa, I., Tani, T., Mukai, A., Sakaguchi, S. and Yasutani, S. (2005) Near-field optical photolithography for high-aspect-ratio patterning using bilayer resist. *Applied Physics Letters*, **86** (20), 201113.

44 Inao, Y., Nakasato, S., Kuroda, R. and Ohtsu, M. (2006) Near-field lithography as prototype nano-fabrication tool. presented at the 32nd International Conference on Micro- and Nano-Engineering, 17–20, September 2006, Barcelona, Spain, Paper 2C-6, Centro Nacional de Microelectronica.

45 Kawazoe, T., Kobayashi, K., Sangu, S., Ohtsu, M. and Neogi, A. (2006) Unique Properties of Optical Near-field and their

Applications to Nanophotonics, in: *Progress in Nano-Electro-Optics V*, Springer-Verlag, Berlin, pp. 109–162.

46 Naruse, M., Kawazoe, T., Sangu, S., Kobayashi, K. and Ohtsu, M. (2006) Optical interconnects based on optical far- and near-field interactions for high-density data broadcasting. *Optics Express*, **14** (1), 306–313.

47 Matsumoto, T., Anzai, Y., Shintani, T., Nakamura, K. and Nishida, T. (2006) Writing 40 nm marks by using a beaked metallic plate near-field optical probe. *Optics Letters*, **31** (2), 259–261.

48 Naruse, M., Yatsui, T., Nomura, W., Hirose, N. and Ohtsu, M. (2005) Hierarchy in optical near-fields and its application to memory retrieval. *Optics Express*, **13** (23), 9265–9271.

49 Ito, H., Nakata, T., Sakaki, K., Ohtsu, M., Lee, K.I. and Jhe, W. (1996) Laser spectroscopy of atoms guided by evanescent waves in micron-sized hollow optical fibers. *Physical Review Letters*, **76** (24), 4500–4503.

50 Totsuka, K., Ito, H., Kawamura, T. and Ohtsu, M. (2002) High spatial resolution atom detector with two-color optical near-fields. *Japanese Journal of Applied Physics*, **41** (3A), 1566–1571.

51 Totsuka, K., Ito, H., Suzuki, K., Yamamoto, K., Ohtsu, M. and Yatsui, T. (2003) A slit-type atom deflector with near-field light. *Applied Physics Letters*, **82** (10), 1616–1618.

52 Takamizawa, A., Ito, H., Yamada, S. and Ohtsu, M. (2004) Observation of cold atom output from an evanescent-light funnel. *Applied Physics Letters*, **85** (10), 1790–1792.

53 Inoue, T. and Hori, H. (2001) Quantization of evanescent electromagnetic waves based on detector modes. *Physical Review A*, **63** (6), 063805.

54 Inoue, T. and Hori, H. (2005) Quantum theory of radiation in optical near-field based on quantization of evanescent electromagnetic waves using detector mode, in: *Progress in Nano-Electro-Optics IV*, Springer-Verlag, Berlin, pp. 127–199.

55 Inoue, T. and Hori, H. (2005) Theory of transmission and dissipation of radiation near a metallic slab based on angular spectrum representation. *IEICE Transactions on Electronics*, **E88-C** (9), 1836–1844.

56 Cho, K. (1991) Nonlocal theory of radiation-matter interaction: Boundary-condition-less treatment of Maxwell equation. *Progress of Theoretical Physics Supplement*, **106**, 225–233.

57 Ishihara, H. and Cho, K. (1993) Nonlocal theory of the third-order nonlinear optical response of confined excitons. *Physical Review B-Condensed Matter*, **48** (11), 7960–7974.

58 Cho, K., Ohfuti, Y. and Arima, K. (1996) Theory of resonant SNOM (scanning near-field optical microscopy): breakdown of the electric dipole selection rule in the reflection mode. *Surface Science*, **363** (1–3), 378–384.

59 Iida, T. and Ishihara, H. (2005) Optical manipulation of nano materials under quantum mechanical resonance conditions. *IEICE Transactions on Electronics*, **E88-C** (9), 1809–1816.

60 Mitsutake, K. and Tsukada, M. (2006) Theoretical study of electron-vibration coupling on carrier transfer in molecular bridges. *e-Journal of Surface Science and Nanotechnology*, **4**, 311–318.

61 Chen, L.C., Yonehama, J., Ueda, T., Hori, H. and Hiraoka, K. (2007) Visible-laser desorption/ionization on gold nanostructures. *Journal of Mass Spectrometry*, **42** (3), 346–353.

62 Chen, L.C., Ueda, T., Sagisaka, M., Hori, H. and Hiraoka, K. (2007) Visible laser desorption/ionization mass spectrometry using gold nanorods. *The Journal of Physical Chemistry*, **111** (6), 2409–2415.

63 Shannon, C.E. (1956) The bandwagon. *IEEE Transactions on Information Theory*, **IT-2** (1), 3.

2
Nanofabrication Principles and Practice
Tadashi Kawazoe and Motoichi Ohtsu

2.1
Adiabatic Nanofabrication

To introduce the concept of nanofabrication by nanophotonics, this section reviews adiabatic nanofabrication processes, such as the use of an optical near-field. This is an example of the quantitative innovation of nanofabrication. Near-field optical chemical vapor deposition (NFO-CVD; Figure 2.1) enables the fabrication of nanometer-scale structures, while precisely controlling their size and position [1–6]. This means that the position can be controlled accurately by regulating the position of the fiber probe that generates the optical near-field. A sharpened ultraviolet (UV) fiber probe is used to guarantee the generation of an optical near-field of sufficiently high efficiency; this is generally fabricated using a pulling/etching technique [7]. NFO-CVD can be used on various materials, including metals, semiconductors, and insulators.

Conventional optical CVD involves two process steps: photodissociation and deposition. For photodissociation to occur, a far-field propagating light must cause the reacting gaseous molecules to resonate and excite the electrons in the molecules from the ground state to an excited electronic state [8, 9]. The Franck–Condon principle holds that this resonance is essential for excitation [8]. The excited electrons then relax to the dissociation channel, and the dissociated Zn atoms are deposited on the substrate surface. Since the electrons in a molecule are excited, this process is adiabatic, and the technique is referred to as adiabatic nanofabrication.

An example of NFO-CVD is the deposition of a Zn dot. Using a sharpened UV fiber probe, gas-phase diethylzinc (DEZn) is dissociated selectively and Zn dots of 20-nm diameter can be fabricated successfully with a 65-nm separation on a sapphire (0001) substrate (see Figures 2.2(a) and (b)) [6]. By changing the reactant molecules during deposition, nanometric Zn and Al dots can be deposited successively on the same sapphire substrate with high precision (see Figure 2.2(c)) [4].

Figure 2.3(a) shows a shear-force image of four Zn dots deposited with irradiation times of 60 (dot 1), 30 (dot 2), 10 (dot 3), and 5 s (dot 4) with a laser output power of 5 µW. The Zn dots were deposited at separations of 300 and 260 nm along the *x*- and

2 Nanofabrication Principles and Practice

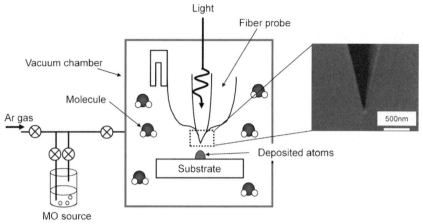

Figure 2.1 Schematic explanation of NFO-CVD.

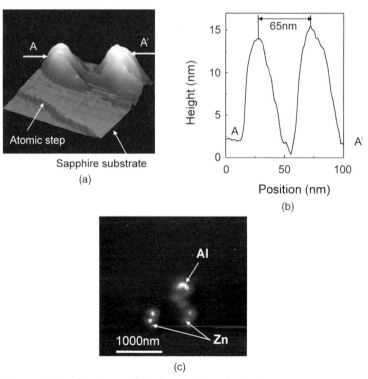

Figure 2.2 Shear-force image of closely spaced dots: (a) Zn dots, (b) cross-sectional profile along the line indicated by arrows A–A' in (a), and (c) Zn and Al dots.

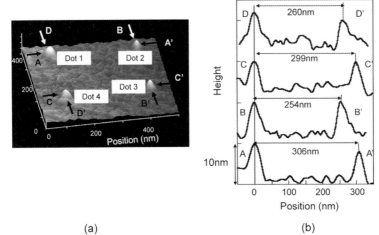

(a)   (b)

Figure 2.3 Deposited Zn dots. (a) Shear-force image. The laser irradiation times for dots 1–4 were 60, 30, 10, and 5 s, respectively. (b) Cross-sectional profiles along the lines indicated by arrows A–A', B–B', C–C', and D–D', respectively.

y-axes, respectively, under servo-control of the fiber probe position. Zn dots as small as 30 nm in diameter can be fabricated, as shown by the cross-sectional profiles in Figure 2.3(b). The figure indicates separations of 306 and 299 nm along the x-axis, and 260 and 254 nm along the y-axis, confirming a high positional accuracy of less than 10 nm. The main source of the residual inaccuracy is the hysteresis of the piezoelectric transducer used for scanning the fiber probe. This can be decreased by carefully selecting the transducer.

Such an adiabatic process has also been used in near-field photolithography, which will be reviewed in Chapter 6.

2.2
Nonadiabatic Nanofabrication

This section describes the nonadiabatic processes involved in optical CVD and photolithography. These methods are responsible for a qualitative innovation in nanofabrication due to the spatially localized nature of optical near-fields.

2.2.1
Nonadiabatic Near-Field Optical Chemical Vapor Deposition

In NFO-CVD, photodissociation can also occur in a nonadiabatic process under nonresonant conditions. The photodissociation of metal organic molecules and the deposition of Zn dots have been achieved using a nonresonant optical near-field (ONF) with a photon energy lower than the energy gap of the electronic state of the molecule [10]. This photochemical reaction is one of the unique ONF phenomena;

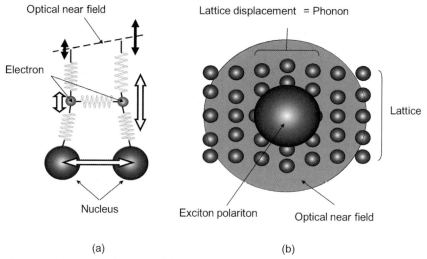

Figure 2.4 Schematic explanations of the excitation of molecular vibration modes by the (a) ONF and (b) exciton-phonon-polariton.

several processes can induce such a reaction. These processes are also applicable to many other photochemical reactions in addition to optical CVD, so it is important to clarify their physical origin. This section describes the relationship between the deposition rate of Zn in NFO-CVD and the incident optical power and photon energy, and gives experimental results based on the features of the ONF and the exciton–phonon-polariton (EPP) model.

The EPP model assumes that the steep spatial gradient of the ONF is able to excite a molecular vibration mode. Figure 2.4 illustrates the excitation of the molecular vibration mode by the ONF and EPP. In the case of an optical far-field propagating light, only the electrons in the molecule respond to the optical electric field with the same phase and intensity due to the uniform intensity of the light in a subwavelength molecule. Therefore, far-field light cannot excite the molecular vibration. In contrast, the ONF intensity is not uniform in a molecule because of its steep spatial gradient. The electrons respond due to this nonuniformity, exciting the molecular vibration modes. The molecule is then polarized as a result of this nonuniform response of the electrons, as shown in Figure 2.4(a). The EPP model has been proposed to describe this excitation process quantitatively. The EPP is a quasiparticle consisting of an exciton–polariton carrying the phonon (lattice vibration) generated by the steep spatial gradient of its optical field, as shown in Figure 2.4(b).

We start by presenting experimental results. The cone angle and apex diameter of the fiber probe used for NFO-CVD were 30° and less than 30 nm, respectively [11]. Since a bare fiber probe without an opaque coating was used for the deposition, far-field light escaped from it in addition to an ONF that was generated at the apex of the probe. This allowed us to investigate the deposition by the far-field light and the ONF simultaneously. The optical power from the fiber probe was measured with a photodiode placed behind the sapphire substrate. The deposited Zn dots were measured using a shear-force microscope with the fiber probe used for deposition. The buffer gas was ultrahigh purity Ar at a pressure of 3 Torr, and the reactant gaseous molecule was DEZn at 100 mTorr and room temperature. A He-Cd laser ($\hbar\omega = 3.81$ eV, where ω is the angular frequency of light) was used as a nearly resonant light source with the absorption band edge E_{abs} (4.13 eV) of DEZn [9]. Ar^+ (2.54 eV) and diode lasers (1.81 eV) were used as nonresonant light sources.

Figure 2.5 shows shear-force topographical images of the sapphire substrate after NFO-CVD using ONFs of 3.81, 2.54, and 1.81 eV. The laser power and irradiation

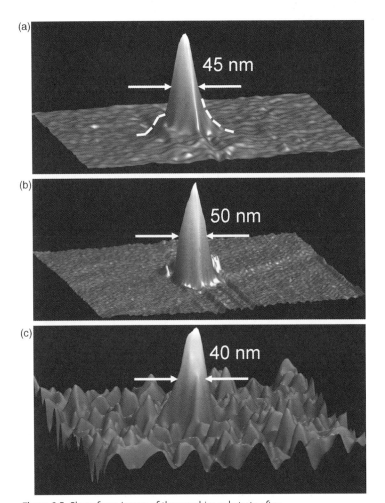

Figure 2.5 Shear-force image of the sapphire substrate after NFO-CVD for photon energies, laser output powers, and deposition times of (a) 3.81 eV, 2.3 µW, and 60 s; (b) 2.53 eV, 360 µW, and 180 s; and (c) 1.81 eV, 1 mW, and 180 s. The scanning areas are all 450 × 450 nm.

time were 2.3 μW, 60 s; 360 μW, 180 s; and 1 mW, 180 s, respectively. While the experimental results of conventional far-field optical CVD show that a Zn film cannot be grown by the nonresonant light ($\hbar\omega < 4.13$ eV) [12], deposited Zn dots were observed on the substrate just below the apex of the fiber probe even with nonresonant light. The composition of the deposited material was confirmed by X-ray photoelectron spectroscopy. Moreover, luminescence was observed from nanometric ZnO dots that were grown by oxidizing the Zn dots fabricated by NFO-CVD [3]. These experimental results imply that the Zn was very pure. In Figure 2.5(a), the photon energy exceeded the dissociation energy ($E_d = 2.26$ eV) of DEZn, and was close to the E_{abs} of DEZn, i.e. $\hbar\omega > E_d$ and $\hbar\omega \approx E_{abs}$ [9]. The topographical image was 45 nm in diameter and 26 nm high, with long tails shown by the dotted curves. These tails represented a Zn layer less than 2 nm thick that was deposited by far-field light leaking from the bare fiber probe. This deposition was possible because DEZn can absorb the light of $\hbar\omega = 3.81$ eV to a small degree. The high peak in the image suggests that ONF enhanced the photodissociation rate as the ONF intensity increased acutely near the apex of the fiber probe. In Figure 2.5(b), the photon energy still exceeded the dissociation energy of DEZn, but it was lower than the absorption edge, i.e. $\hbar\omega > E_d$ and $\hbar\omega > E_{abs}$ [9, 13]. The image was 50 nm in diameter and 24 nm high. Although a high intensity far-field light leaked from the bare fiber probe, it did not deposit a Zn layer. Thus, there was no tail at the foot of the peak. This confirmed that the photodissociation of DEZn and Zn deposition occurred due to an ONF of $\hbar\omega = 2.54$ eV. Figure 2.5(c) shows the result for $\hbar\omega < E_d$ and $\hbar\omega < E_{abs}$. A Zn dot was successfully deposited even at such a low photon energy. The topographical image was 40 nm in diameter and 2.5 nm high. It appears that these depositions were due specifically to the ONF because the Zn dots were deposited on the substrate just below the apex of the fiber probe even though a high-intensity far-field light leaked from the bare fiber probe.

To analyze this novel photodissociation process quantitatively, the relationship between the photon flux I and the deposition rate R of Zn was examined, as shown in Figure 2.6. For $\hbar\omega = 3.81$ eV, R was proportional to I. For $\hbar\omega = 2.54$ and 1.81 eV, higher-order dependencies appeared and were fitted by the third-order polynomial $R = a \cdot I + b \cdot I^2 + c \cdot I^3$. For $\hbar\omega = 3.81$ eV, $a_{3.81} = 5.0 \times 10^{-6}$ and $b_{3.81} = c_{3.81} = 0$. For $\hbar\omega = 2.54$ eV, $a_{2.54} = 4.1 \times 10^{-12}$, $b_{2.54} = 2.1 \times 10^{-27}$, and $c_{2.54} = 1.5 \times 10^{-42}$. For $\hbar\omega = 1.81$ eV, $a_{1.81} = 0$, $b_{1.81} = 4.2 \times 10^{-29}$, and $c_{1.81} = 3.0 \times 10^{-44}$. These values were used to investigate the physical origin of nonadiabatic NFO-CVD as described below.

Figure 2.7 shows the potential curves of an electron in a DEZn molecular orbit drawn as a function of the internuclear distance of the C—Zn bond, which is involved in photodissociation [9]. The relevant energy levels of the molecular vibration mode are indicated by the horizontal broken lines in each potential curve. When far-field light is used, photoabsorption (indicated by the white arrow in this figure) triggers the dissociation of DEZn [14]. In contrast, when nonresonant ONF is used, there are three possible origins of photodissociation [10]. They are: (1) a multiphoton absorption process, (2) a multistep transition process via the intermediate energy level induced by the fiber probe, and (3) a multistep transition via an excited state of the

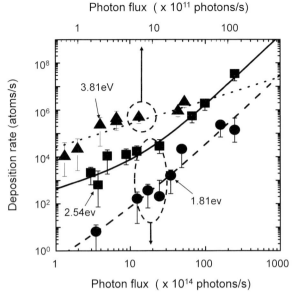

Figure 2.6 The optical-power dependency of the Zn deposition rate. The dotted, solid, and dashed curves represent the calculated results, and the fitted experimental results.

molecular vibration mode. Origin (1) is negligible because the optical power density in the experiment was less than 10 kW/cm², which is too low for multiphoton absorption [15]. Origin (2) is also negligible because NFO-CVD was observed for the light in the ultraviolet (UV) to the near-infrared region, even though DEZn does not

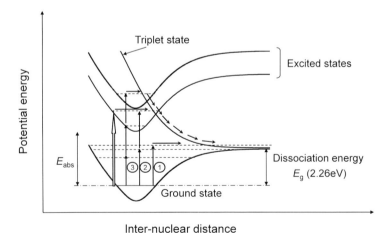

Figure 2.7 Schematic diagram of the potential curves of an electron in DEZn molecular orbits. The relevant energy levels of the molecular vibration modes are indicated by the horizontal dashed lines.

have any relevant energy levels for such a broad spectral region. As a result, these experimental results strongly supported origin (3), indicating that the three multi-step excitation processes in Figure 2.7, labeled ①, ②, and ③, contributed to the photodissociation.

As the second step of this discussion, we review the EPP model to evaluate these calculations. The EPP model was proposed to describe the ONF generated at the nanometric probe tip [16]. An ONF is a highly mixed state with material excitation. In particular, electronic excitation near the probe tip driven by the incident photons causes mode–mode or anharmonic couplings of phonons. They are considered to be renormalized phonons that allow multiphonon transfer from the tip to a molecule simultaneously. The model Hamiltonian for the probe tip can be diagonalized using conventional theory [17], and represented by a quasiparticle: $H = \sum_p \hbar\omega(p)\xi_p^\dagger \xi_p$. Here, the creation (annihilation) operator for the EPP and the angular frequency are denoted $\xi_p^\dagger(\xi_p)$ and $\omega(p)$, respectively. Therefore, in this model, a molecule near the probe tip does not absorb simple photons; instead, it absorbs EPP. As a result, the EPP energy is transferred to the molecule, which excites molecular vibrations or induces electronic transitions.

To illustrate the dissociation probability of a molecule, the transition from the initial to the final states can be formulated by the conventional perturbation method for the interaction Hamiltonian. It is given by the multipolar quantum electrodynamics (QED) Hamiltonian in the dipole approximation [18] for an ONF–molecule interaction, and is expressed as

$$H_{\text{int}} = -\int \boldsymbol{\mu}(\boldsymbol{r}) \cdot \boldsymbol{D}^\perp(\boldsymbol{r}) d^3 r$$

where

$$\boldsymbol{D}^\perp(\boldsymbol{r}) = i\sum_p \left(\frac{2\pi\hbar\omega_p}{V}\right)^{1/2} \boldsymbol{\varepsilon}_p \left[a_p \exp(i\boldsymbol{p}\boldsymbol{r}) - a_p^\dagger \exp(-i\boldsymbol{p}\boldsymbol{r})\right]$$

Here, $\boldsymbol{\mu}(\boldsymbol{r})$ and $\boldsymbol{D}^\perp(\boldsymbol{r})$ denote the electric dipole operator and the electric displacement vector at position $\boldsymbol{r}$, respectively. The polarization unit vector of a photon is designated as $\boldsymbol{\varepsilon}_p$. Rewriting the photon operators ($a_p^\dagger$, a_p) in terms of the exciton-phonon-polariton operators ($\xi_p^\dagger$, ξ_p), and noting that the electric dipole operator consists of two components (electronic and vibrational), the interaction Hamiltonian is expressed in terms of the EPP as

$$H_{\text{int}} = i\{\mu^{\text{el}}(e + e^\dagger) + \mu^{\text{nucl}}(v + v^\dagger)\} \sum_p \sqrt{\frac{2\pi\hbar\omega_p}{V}} \{v_p v'_p(\xi_p + \xi_p^\dagger)\} e^{i\boldsymbol{p}\cdot\boldsymbol{r}}$$

Here, μ^{el} and μ^{nucl} are the electronic and vibrational dipole moments, respectively, and the creation (annihilation) operators of the electronic and vibrational excitations are denoted $e^\dagger(e)$ and $v^\dagger(v)$ respectively. The incident photon angular frequency and transformation coefficients are ω_p and $v_p(v'_p)$, respectively. Then, the transition probability of one-, two-, and three-step excitation (labeled ①, ②, and ③ in Figure 2.7, and denoted by the corresponding final states as $|f_{\text{first}}\rangle$, $|f_{\text{second}}\rangle$, and $|f_{\text{third}}\rangle$) can be written as:

$$P_{\text{first}}(\omega_p) = \frac{2\pi}{\hbar} |\langle f_{\text{first}}|H_{\text{int}}|i\rangle|^2 = \frac{(2\pi)^2}{\hbar d} v_p^2 v_p'^2 u_p'^2 (\mu^{\text{nucl}})^2 (\hbar\omega_p) I_0(\omega_p),$$

$$P_{\text{second}}(\omega_p) = \frac{2\pi}{\hbar} |\langle f_{\text{second}}|H_{\text{int}}|i\rangle|^2$$

$$= \frac{(2\pi)^3}{\hbar d^2} \frac{v_p^4 v_p'^6 u_p'^2}{|\hbar\omega(p)-(E_a-E_i+i\gamma_m)|^2} (\mu^{\text{el}})^2 (\mu^{\text{nucl}})^2 (\hbar\omega_p)^2 I_0^2(\omega_p),$$

$$P_{\text{third}}(\omega_p) = \frac{2\pi}{\hbar} |\langle f_{\text{third}}|H_{\text{int}}|i\rangle|^3$$

$$= \frac{(2\pi)^4}{\hbar d^3} \frac{v_p^6 v_p'^{10} u_p'^2}{|\hbar\omega(p)-(E_a-E_i+i\gamma_m)|^2 |\hbar\omega(p)-(E_{ex}-E_g+i\gamma_m)|^2}$$

$$\times (\mu^{\text{el}})^4 (\mu^{\text{nucl}})^2 (\hbar\omega_p)^3 I_0^3(\omega_p)$$

where d, u'_p, and $I_0(w_p)$ represent the probe tip size, the transformation coefficient, and the incident light intensity, respectively. To ensure the energy conservation in each transition probability, the following initial and three final states of the system were prepared, each consisting of the fiber probe and a molecule:

$$|i\rangle = |\text{probe}\rangle \otimes |E_g; \text{el}\rangle \otimes |E_i; \text{vib}\rangle$$
$$|f_{\text{first}}\rangle = |\text{probe}\rangle \otimes |E_g; \text{el}\rangle \otimes |E_a; \text{vib}\rangle$$
$$|f_{\text{second}}\rangle = |\text{probe}\rangle \otimes |E_{ex}; \text{el}\rangle \otimes |E_b; \text{vib}\rangle$$
$$|f_{\text{third}}\rangle = |\text{probe}\rangle \otimes |E_{ex'}; \text{el}\rangle \otimes |E_c; \text{vib}\rangle$$

where $|\text{probe}\rangle$, $|E_\alpha; \text{el}\rangle$, and $|E_\beta; \text{vib}\rangle$ represent a probe state, molecular electronic state, and vibrational state, respectively. In addition, $E_\alpha (\alpha = g, ex, ex')$ and $E_\beta (\beta = i, a, b, c)$ represent the molecular electronic and vibrational energies, respectively, as shown schematically in Figure 2.7. The γ_m and γ'_m are the spectral linewidth of the vibrational and electronic states, respectively. It follows that these near-resonant transition probabilities have the following ratio:

$$\frac{P_{\text{second}}(\omega_p)/I_0^2(\omega_p)}{P_{\text{first}}(\omega_p)/I_0(\omega_p)} = \frac{P_{\text{third}}(\omega_p)/I_0^3(\omega_p)}{P_{\text{second}}(\omega_p)/I_0^2(\omega_p)} = \frac{\hbar}{2\pi} \frac{P_{\text{first}}(\omega_p)}{|\gamma_m|^2 I_0(\omega_p)} \left(\frac{v_p'^2}{u_p'^2}\right) \left(\frac{\mu^{\text{el}}}{\mu^{\text{nucl}}}\right)^2$$

Here, we assume that $\gamma_m = \gamma'_m$, for simplicity. Using this ratio, the experimental intensity dependence of the deposition rate is analyzed to clarify the origin (3). For $\hbar\omega = 2.54 \text{ eV}$, all the processes (①, ②, and ③) depicted in Figure 2.7 are possible, because $\hbar\omega > E_d$ even though $\hbar\omega < E_{\text{abs}}$. Fitting the experimental value of $P_{\text{first}}(\omega_{2.54}) = a_{2.54} I_0(\omega_{2.54}) = 10^2 \text{events/s}$ with reasonable values of $\mu^{\text{nucl}} = 1$ Debye, $\mu^{\text{el}} = 10^{-3}$ Debye, $\gamma_m = 10^{-1}$ eV, $v_p'^2/u_p'^2 = 0.01$, and $d = 30$ nm, we obtain the following value of the ratio:

$$\frac{P_{\text{second}}(\omega_{2.54})/I_0^2(\omega_{2.54})}{P_{\text{first}}(\omega_{2.54})/I_0(\omega_{2.54})} = \frac{P_{\text{third}}(\omega_{2.54})/I_0^3(\omega_{2.54})}{P_{\text{second}}(\omega_{2.54})/I_0^2(\omega_{2.54})}$$

$$= \frac{\hbar}{2\pi} \frac{P_{\text{first}}(\omega_{2.54})}{|\gamma_m|^2 I_0(\omega_{2.54})} \left(\frac{v_p'^2}{u_p'^2}\right) \left(\frac{\mu^{\text{el}}}{\mu^{\text{nucl}}}\right)^2 \simeq 10^{-15}$$

which is in good agreement with the experimental values $b_{2.54}/a_{2.54} \simeq c_{2.54}/b_{2.54} \simeq 10^{-15}$. For $\hbar\omega = 1.81$ eV, dissociation occurs via either ② or ③ shown in Figure 2.7 because $\hbar\omega < E_d$ (E_{abs}). The ratio can be evaluated as

$$\frac{P_{third}(\omega_{1.81})/I_0^3(\omega_{1.81})}{P_{second}(\omega_{1.81})/I_0^2(\omega_{1.81})} = \frac{\hbar}{2\pi} \frac{P_{first}(\omega_{1.81})}{|\gamma_m|^2 I_0(\omega_{1.81})} \left(\frac{v_p'^2}{u_p'^2}\right) \left(\frac{\mu^{el}}{\mu^{nucl}}\right)^2 \simeq 10^{-15}$$

which is also in good agreement with the experimental value $c_{1.81}/b_{1.81} \simeq 10^{-15}$. The experimental value of $P_{first}(\omega_{1.81}) \simeq a_{2.54} I_0(\omega_{2.54}) = 10^2$ events/s is used as a theoretical estimate because both transitions by the photon energies of 1.84 and 2.54 eV are attributable to the coupling between phonons in the probe and molecular vibrations. The overall agreement between the theoretical and experimental results suggests that the EPP model provides a way to understand the physical origin of the nonadiabatic photodissociation process. For $\hbar\omega = 3.81$ eV, the direct absorption by the electronic state is much stronger than other cases because the light is nearly resonant with DEZn. This is why no higher-order power dependence of the deposition rate was observed.

As an example of nonadiabatic NFO-CVD, the zinc-bis(acetylacetonate) (Zn(acac)$_2$) molecule was dissociated. This molecule has never been used for conventional optical CVD due to its low optical activity. With NFO-CVD, however, the ONF can activate the molecule nonadiabatically for dissociation and deposition. Figure 2.8(a) shows a shear-force topographical image of Zn deposited on a sapphire substrate. The laser power and irradiation time were 1 mW and 15 s, respectively. The Zn dot was 70 nm in diameter and 24 nm high [19, 20]. The chemical stability of Zn(acac)$_2$ maintains a clean substrate surface, which is an advantage in fabricating an isolated nanostructure. Figure 2.8(b) shows a shear-force topographical image of a deposited Zn dot, 5 nm in diameter and 0.3 nm high, that is among the smallest ever fabricated using NFO-CVD. The deposition conditions were gaseous Zn(acac)$_2$ molecules with a pressure of 70 mTorr in the CVD chamber, and a laser wavelength, power, and irradiation time of 457 nm, 65 μW, and 30 s, respectively.

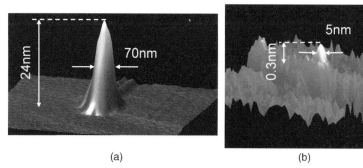

Figure 2.8 Shear-force images after NFO-CVD using Zn(acac)$_2$ with a 457-nm wavelength light source: (a) deposited Zn dot with a diameter of 70 nm and a height of 24 nm and (b) deposited Zn dot with a diameter of 5 nm and a height of 0.3 nm.

2.2.2
Nonadiabatic Near-Field Photolithography

A nonadiabatic photochemical process has been applied to near-field photolithography [20, 21]. This section describes the dependence of the fabricated pattern on the exposure.

Several methods of nanofabrication have been developed to meet the increasing requirements for mass production of photonic and electronic devices [22–25]. Compared to these, nonadiabatic photolithography has the potential for wider use because it permits the use of conventional photolithographic components and systems for nanofabrication beyond the diffraction limit of light. However, the greatest advantage of nonadiabatic photolithography is that it is free from the coherency and polarization problems that are intrinsic to the wave properties of incident light.

Figure 2.9(a) shows a schematic diagram of a photomask and a photoresist that was spin-coated on a Si substrate. The gap between the photoresist and photomask was controlled to less than 20 nm by mechanical compression or air pressure (Figure 2.9(b)). In conventional adiabatic photolithography, the photoresist is exposed by the ONF and the far-field light from the UV light source. Thus, the exposed region of the photoresist spreads across the slit width due to the diffraction of light, as shown in Figure 2.9(c) [26, 27]. Nonadiabatic photolithography, on the other hand, uses visible light, to which the photoresist is insensitive. Therefore, the photoresist is exposed only by the ONF generated at the slit of the photomask (Figure 2.9 (d)) [21]. This selective exposure increases the resolution.

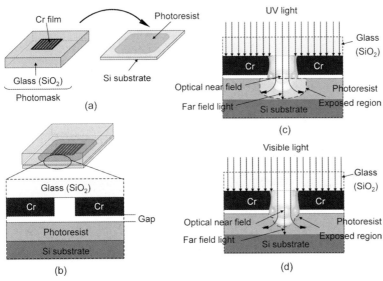

Figure 2.9 Schematic diagrams of the photomask and photoresist: (a) and (b) setup for the exposure process, (c) and (d) explanations of the adiabatic and nonadiabatic exposure processes.

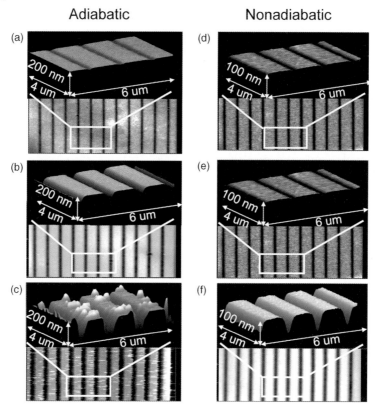

Figure 2.10 Developed surfaces of the photoresist: (a)–(c) adiabatic photolithography and (d)–(f) nonadiabatic photolithography.

Figure 2.10 shows atomic force microscopy (AFM) images of the photoresist after exposure. A Xe lamp with two optical band-pass filters was used as the light source in the exposure process. The photomask was made of Cr film with 1.5-μm wide lines and optically transparent 150-nm wide spaces on a 2-mm thick SiO_2 substrate. The photoresist was OFPR-800 (Tokyo-Ohka Co., Ltd., Japan), which has high optical sensitivity for the g-line of a Hg lamp (wavelength of 436 nm) but is insensitive to wavelengths longer than 500 nm. It was spin-coated onto the Si substrate to a thickness of 500 nm. For adiabatic near-field lithography, a filtered Xe lamp was used with a center wavelength of 400 nm and full-width at half-maximum (FWHM) of 40 nm. The optical power density at the front surface of the photomask was 1 W/cm². The transferred patterns on the photoresist surface after exposure and development are shown in Figures 2.10(a)–(c) for exposure times of 1, 2, and 5 s, respectively. In this case, the photoresist was exposed by both the ONF and the far-field light, so the exposed region expanded with increasing optical intensity. At the optimal exposure times of 1–2 s, the groove width was about 200 nm, which is slightly wider than the space of the photomask patterns. However, the pattern width rapidly increased to

500 nm when the exposure time was increased to only twice the optimum value (Figure 2.10(c)). In addition, the resist surface became rough due to the diffracted far-field light.

For nonadiabatic near-field lithography, a filtered Xe lamp was used with a center wavelength of 550 nm and a FWHM of 40 nm. The optical power density at the front surface of the photomask was 1 W/cm^2. The transferred patterns on the photoresist surface are shown in Figures 2.10(d)–(f) after exposure and development for exposure times of 60, 120, and 480 s, respectively. In this case, the photoresist was exposed mainly by the ONF, and only slightly by the far-field light. At the optimal exposure time of 60–120 s, the groove width was about 150 nm, which corresponded to the space width of the photomask. The groove depth was also regulated by the spatial distribution of the ONF. The widths and depths of the grooves were only slightly greater, even at 480 s (Figure 2.10(f)) although this exposure time is as much as four times the optimal value. Independence of the fabricated pattern size on the exposure time is due to the very low sensitivity to the far-field light. Note that the resist surface remained clean.

Figure 2.11 shows the exposure time dependence of the developed depth of the photoresist for adiabatic and nonadiabatic photolithography under the same exposure conditions as Figure 2.10. The depths were measured by AFM after development. The developed depth increased linearly under adiabatic exposure. For nonadiabatic exposure, on the other hand, the developed depths consisted of two parts. One part agreed with the fitted curve based on the assumption of a saturation depth of 10 nm, shown as region **A** in Figure 2.11. The other part agreed with the fitted curve based on the assumptions of a 400-nm saturation depth and a threshold of 350-s exposure time, shown as region **B** in Figure 2.11. It was assumed that region **A** corresponds to the region exposed by the ONF at the boundary between the photomask and photoresist since the ONF is localized at the nanometric surface.

To investigate region **B**, the optical field intensity was calculated using the finite-difference time-domain (FDTD) method. Since the coupling between the exciton–polariton and phonon is caused by the field gradient force in a nonadiabatic photochemical process [28], the exposure rate is proportional to the product of the

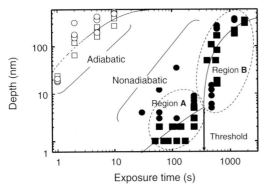

Figure 2.11 Dependence of the developed depth on the exposure time.

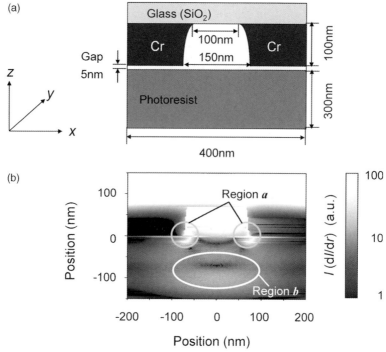

Figure 2.12 Simulation by the FDTD method: (a) simulation model and (b) calculated spatial distribution of $I(r) \cdot (dI(r)/dr)$.

field intensity $I(r)$ and its spatial gradient. Figures 2.12(a) and (b) show the FDTD simulated model and the simulated intensity distribution of $I(r)\cdot(dI(r)/dr)$, respectively, where is $dI(r)/dr = \sqrt{\left\{\frac{\partial I(x,y,z)}{\partial x}\right\}^2 + \left\{\frac{\partial I(x,y,z)}{\partial y}\right\}^2 + \left\{\frac{\partial I(x,y,z)}{\partial z}\right\}^2}$ is the spatial gradient of $I(r)$. A cell size of 2 nm and a light source wavelength of 550 nm were used in the FDTD simulation. The light source had a linear polarization of 45° in the x–y plane. The boundary conditions for the x- and y-axes were periodic, and a perfectly matched layer boundary condition for the z-axis was used. The refractive index of the photoresist was 1.9, which was estimated from the experimental results. The value of $I(r)\cdot(dI(r)/dr)$ had a local maximum near the edges of the Cr mask in region *a*, which corresponds to region **A** in Figure 2.11. It should be noted that the $I(r)\cdot(dI(r)/dr)$ also had a local maximum in region *b*, which is under the surface inside the photoresist. As the exposure time increased, region *b* grew and eventually reached the surface of the photoresist. Thus, the developed region suddenly increased at the threshold exposure time. This progress of exposure explains the dependence of the developed depth in region **B** (Figure 2.11).

Figure 2.13 shows other experimental results to clarify the details of this exposure process. In this experiment, a photomask of narrower 200-nm lines and 100-nm

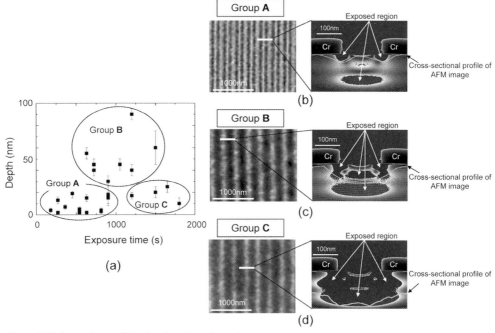

Figure 2.13 Dependence of the developed depth on the exposure time.

spaces was used. The depth of the patterns transferred onto the photoresist depended on the exposure times, which were classified in three groups indicated as **A**, **B**, and **C** in Figure 2.13(a). Figures 2.13(b), (c), and (d) show the typical patterns for groups **A**, **B** and **C**, respectively, which were measured by AFM along with their cross-sectional profiles. Calculated cross-sectional profiles $\int I(r) \cdot (dI(r)/dr)dt$ are also shown in these figures. The estimated regions exposed by the nonadiabatic photochemical process agreed well with the cross-sectional profiles of the developed photoresist. This agreement supports the premise of a nonadiabatic photochemical process. The exposure processes in each group can be explained as follows. In Figure 2.13(b) (group **A**), exposed regions appear near the edges of the Cr mask and independently inside the photoresist. Therefore, the exposed region is shallow, and two grooves with a width of 70 nm (FWHM) are fabricated at each edge of the photomask. In Figure 2.13 (c) (group **B**), exposed regions near the edges of the Cr mask and inside the photoresist connect to each other to form one deep groove. Thus, the width of the groove increases to 200 nm. In Figure 2.13(d) (group **C**), the exposed region connects to the neighboring exposed region, resulting in a shallower groove.

Nonadiabatic photolithography has been used in the fabrication of more complicated patterns. Figure 2.14 shows AFM images of the photoresist surface after exposure and development. These patterns were fabricated on TDMR-AR87 photoresist (Tokyo-Ohka Kogyo Co., Japan) using a linearly polarized g-line light. TDMR-AR87 is the photoresist for an i-line (365-nm) light source that is insensitive to the g-line. Reproducible transfers of two-dimensional arrays of circles and T-shapes were

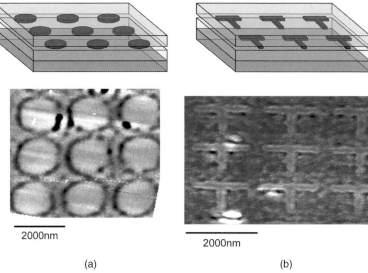

Figure 2.14 AFM images of nonadiabatic lithography using TDMR-AR87 photoresist: (a) exposed by the polarized g-line of a Hg lamp for 10 s with a circle-shaped photomask and (b) developed after a 40-s exposure by the g-line of a Hg lamp for T-shaped arrays.

successful, free from the interference of the polarized light that limits the resolution of conventional adiabatic lithography.

The exposure of an optically inactive electron beam resist (EB: ZEP-520: ZEON Co. Ltd., Japan) has also been tested using nonadiabatic photolithography. Figure 2.15 shows an AFM image of the developed EB resist surfaces after exposure. The light

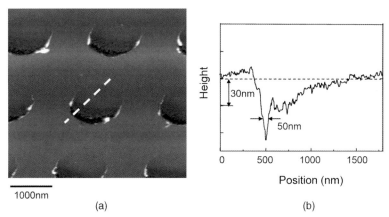

Figure 2.15 An electron-beam resist exposed for 5 min using a Q-switched laser (wavelength of 355 nm) and a circular-shaped array photomask: (a) AFM image and (b) cross-sectional profile along the dashed line in (a).

source was the third harmonic of a Q-switched Nd:YAG laser and the exposure time was 5 min. The pulse energy density, pulse duration, and repetition frequency were $250\,\mu J/cm^2$, 10 ns, and 20 Hz, respectively. The patterning was successful even though the EB resist is insensitive to the far-field propagating light. The developed pattern of 50-nm structures had a pit depth of 70 nm, which is sufficiently deep for the next process such as substrate etching. Since the EB resist has the advantages of homogeneity and a smooth surface, this demonstrates the possibility of using a homogeneous and smooth surface material as a photoresist, even if it is optically inactive.

References

1 Polonski, V.V., Yamamoto, Y., Kourogi, M., Fukuda, H. and Ohtsu, M. (1999) Nanometric patterning of zinc by optical near-field photochemical vapour deposition. *Journal of Microscopy*, **194** (2–3), 545–551.

2 Yamamoto, Y., Kourogi, M., Ohtsu, M., Polonski, V.V. and Lee, G.H. (2000) Fabrication of nanometric zinc pattern with photodissociated gas-phase diethylzinc by optical near-field. *Applied Physics Letters*, **76** (16), 2173–2175.

3 Yatsui, T., Kawazoe, T., Ueda, M., Yamamoto, Y., Kourogi, M. and Ohtsu, M. (2002) Fabrication of nanometric single zinc and zinc oxide dots by the selective photodissociation of adsorption-phase diethylzinc using a nonresonant optical field. *Applied Physics Letters*, **81** (19), 3651–3653.

4 Yamamoto, Y., Kourogi, M., Ohtsu, M., Lee, G.H. and Kawazoe, T. (2002) Lateral integration of Zn and Al dots with nanometer-scale precision by near-field optical chemical vapor deposition using a sharpened optical fiber probe. *IEICE Transactions on Electronics*, **E85-C** (12), 2081–2085.

5 Yatsui, T., Takubo, S., Lim, J., Nomura, W., Kourogi, M. and Ohtsu, M. (2003) Regulating the size and position of deposited Zn nanoparticles by optical near-field desorption using size-dependent resonance. *Applied Physics Letters*, **83** (9), 1716–1718.

6 Lim, J., Yatsui, T. and Ohtsu, M. (2005) Observation of size-dependent resonance of near-field coupling between a deposited Zn dot and the probe apex during near-field optical chemical vapor deposition. *IEICE Transactions on Electronics*, **E88-C**, (9), 1832–1835.

7 Mononobe, S. (1998) *Near-Field Nano/Atom Optics and Technology*, Springer-Verlag, Berlin, pp. 31–69.

8 Calvert, J.G. and Pitts, J.N. Jr (1966) *Photochemistry*, Wiley, New York.

9 Jackson, R.L. (1992) Vibrational energy of the monoalkyl zinc product formed in the photodissociation of dimethyl zinc, diethyl zinc, and dipropyl zinc. *Journal of Chemical Physics*, **96** (8), 5938–5951.

10 Kawazoe, T., Yamamoto, Y. and Ohtsu, M. (2001) Fabrication of a nanometric Zn dot by nonresonant near-field optical chemical-vapor deposition. *Applied Physics Letters*, **79** (8), 1184–1186.

11 See for example, M. Ohtsu (ed.), (1998) *Near-Field Nano/Atom Optics and Technology*, Springer-Verlag, Berlin.

12 Shimizu, M., Kamei, H., Tanizawa, M., Shiosaki, T. and Kawabata, A. (1988) Low temperature growth of ZnO film by photo-MOCVD. *Journal of Crystal Growth*, **89** (4), 365–370.

13 Jackson, R.L. (1989) Metal-alkyl bond dissociation energies for $(CH_3)_2Zn$, $(C_2H_5)Zn$, $(CH_3)_2Cd$, and $(CH_3)_2Hg$. *Chemical Physics Letters*, **163** (4–5), 315–322.

14 Okabe, H. (1978) *Photochemistry of Small Molecules*, John Wiley & Sons Inc., New York.

15 Yariv, A. (1985) *Introduction to Optical Electronics*, 3rd edn, Holt, Rinehart and Winstone, Inc., Orlando.

16 Kobayashi, K., Sangu, S., Ito, H. and Ohtsu, M. (2001) Near-field optical potential for a neutral atom. *Physical Review A*, **63** (1), 013806.

17 Hopfield, J.J. (1958) Theory of the Contribution of Excitons to the Complex Dielectric Constant of Crystals. *Physical Review*, **112** (5), 1555–1567.

18 Craig, D.P. and Thirunamachandran, T. (1998) *Molecular Quantum Electrodynamics*, Dover Publications, Inc., New York.

19 Kawazoe, T., Kobayashi, K. and Ohtsu, M. (2006) Near-field optical chemical vapor deposition using $Zn(acac)_2$ with a non-adiabatic photochemical process. *Applied Physics B-Lasers and Optics*, **84** (1–2), 247–251.

20 Kawazoe, T. and Ohtsu, M. (2004) Adiabatic and nonadiabatic nanofabrication by localized optical near-fields. *Proceedings of SPIE*, **5339**, 619–630.

21 Yonemitsu, H., Kawazoe, T., Kobayashi, K. and Ohtsu, M. (2007) Nonadiabatic photochemical reaction and application to photolithography. *Journal of Luminescence*, **122–123**, 230–233.

22 Leonard, D., Krishnamurthy, M., Reaves, C.M., Denbaars, S.P. and Petroff, P.M. (1993) Direct formation of quantum-sized dots from uniform coherent islands of InGaAs on GaAs surfaces. *Applied Physics Letters*, **63** (23), 3203–3205.

23 Tsutsui, T., Kawasaki, K., Mochizuki, M. and Matsubara, T. (1999) Site controlled metal and semiconductor quantum dots on epitaxial fluoride films. *Microelectronic Engineering*, **47** (1–4), 135–137.

24 Kohmoto, S., Nakamura, H., Ishikawa, T. and Asakawa, K. (1999) Site-controlled self-organization of individual InAs quantum dots by scanning tunneling probe-assisted nanolithography. *Applied Physics Letters*, **75** (22), 3488–3490.

25 Kuramochi, E., Temmyo, J., Tamamura, T. and Kamada, H. (1997) Perfect spatial ordering of self-organized InGaAs/AlGaAs box-like structure array on GaAs (311)B substrate with silicon nitride dot array. *Applied Physics Letters*, **71** (12), 1655–1657.

26 Naya, M., Tsuruma, I., Tani, T., Mukai, A., Sakaguchi, S. and Yasunami, S. (2005) Near-field optical photolithography for high-aspect-ratio patterning using bilayer resist. *Applied Physics Letters*, **86** (20), 20113.

27 Inao, Y., Nakasato, S., Kuroda, R. and Ohtsu, M. (2007) Near-field lithography as prototype nano-fabrication tool. *Microelectronic Engineering*, **84** (5–8), 705–710.

28 Kawazoe, T., Kobayashi, K., Takubo, S. and Ohtsu, M. (2005) Nonadiabatic photodissociation process using an optical near-field. *Journal of Chemical Physics*, **122** (2), 024715.

3
Nanofabrications by Self-Organization and Other Related Technologies
Takashi Yatsui, Wataru Nomura, Kazuya Hirata, Yoshinori Tabata, and Motoichi Ohtsu

3.1
Introduction

Progress in DRAM technology requires improved lithography. It is estimated that the technology nodes should be down to 16 nm by the year 2019 [1]. Recent improvement of immersion lithography using excimer laser (wavelength of 193 and 157 nm) has realized the technology node as small as 90 nm. Further decrease in the node is expected using extreme ultraviolet (EUV) light sources with a wavelength of 13.5 nm. However, the resolution of the linewidth is limited by the diffraction limit of the light. Furthermore, continued innovation for transistor scaling is required to manage power density and heat dissipation.

To overcome these difficulties, we have proposed nanometer-scale photonic integrated circuits (i.e. nanophotonic ICs) [2]. These devices consist of nanometer-scale dots, and an optical near-field is used as the signal carrier. Since an optical near-field is free from the diffraction of light due to its size-dependent localization and size-dependent resonance features, nanophotonics enables the fabrication, operation, and integration of nanometric devices.

In this chapter, we review the optical near-field phenomena and their applications to realize the nanophotonic device. To realize nanometer-scale controllability in size and position, we demonstrate the feasibility of nanometer-scale chemical vapor deposition using optical near-field techniques (see Section 3.2). Based on which, the probeless fabrication method for mass production is also demonstrated in Section 3.3. As a method for mass production, Section 3.4 reviews near-field imprint lithography. Section 3.5 reviews nonadiabatic photochemical reactions to be used to realize ultraflat surfaces based on self-assembling.

3.2
Near-Field Optical Chemical Vapor Deposition

As an introduction to nanophotonic fabrication, near-field optical chemical vapor deposition (NFO-CVD) is reviewed; that is, the use of an optical near-field can

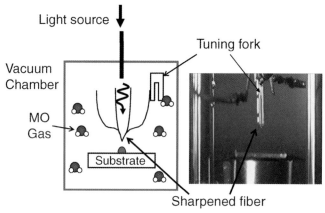

Figure 3.1 Schematic explanation of NFO-CVD.

realize nanoscale fabrication and quantitative innovations are demonstrated. NFO-CVD (Figure 3.1) has been developed, and enables the fabrication of nanometer-scale structures, while controlling their size and position precisely [3–8]. That is, the position can be controlled accurately by regulating the position of the fiber probe used to generate the optical near-field. To guarantee the generation of an optical near-field with sufficiently high efficiency, a sharpened UV fiber probe was used. The sharpened UV fiber probe was fabricated using a pulling/etching technique, in which the fiber was pulled under the illumination of a CO_2 laser (Figure 3.2(a)) and etched with buffered hydrofluoric acid (Figure 3.2(b)) [9]. Under the uncoated condition, the diameter of the sharpened probe tip remained sufficiently small (Figure 3.2(d)), which enabled high-resolution position control and *in situ* shear-force topographic imaging of the deposited nanometer-scale structures. Since the deposition time was sufficiently short, the deposition of metal on the fiber probe and the resultant decrease in the throughput of optical near-field generation were negligible. The separation between the fiber probe and sapphire (0001) substrate was kept within a few nanometers by shear-force feedback control. To realize the shear-force feedback control, the fiber probe was glued to the tuning fork (Figure 3.1). Immediately after the nanodots were deposited, their sizes and shapes were measured via *in situ* vacuum shear-force microscopy [10], using the same probe as used for deposition. Due to the photochemical reaction between the reactant molecules and the optical near-field generated at the tip of an optical fiber probe, NFO-CVD is applicable to various materials, including metals, semiconductors, and insulators.

Conventional optical CVD method uses a light source that resonates the absorption band of a metalorganic (MO) vapor and has a photon energy that exceeds the dissociation energy [11]. Therefore, it utilizes a two-step process: gas-phase photodissociation and subsequent adsorption. In this process, resonant photons excite molecules from the ground state to the excited electronic state and the excited molecules relax to the dissociation channel, and then the dissociated metallic atoms

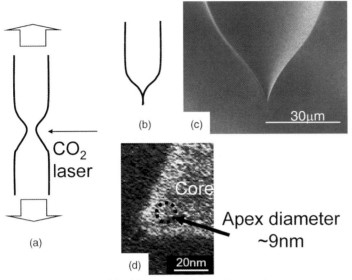

Figure 3.2 Schematic of the fabrication of the UV-sharpened probe. (a) Pulling the fiber. (b) Sharpened fiber using buffered hydrogen fluoride solution. (c) SEM picture of the sharpened fiber. (d) Magnified SEM image of the apex of the fiber.

adsorb to the substrate [12]. However, it was found that the dissociated MO molecules migrate on the substrate before adsorption, which limits the minimum lateral size of the deposited dots (Figure 3.3(a)). A promising method for avoiding this migration is dissociation and deposition in the adsorption phase (Figure 3.3(b)) [13].

An example of NFO-CVD is the deposition of a Zn dot. Since the absorption band-edge energy of gas-phase diethylzinc (DEZn) is 4.6 eV ($\lambda = 270$ nm) [11], a He–Cd laser (3.81 eV, $\lambda = 325$ nm) was used as the light source for the deposition of Zn; it is nonresonant to gas-phase DEZn. However, a red-shift occurs in the absorption spectrum of DEZn with respect to that in the gas-phase; that is, it resonates the adsorption phase DEZn. The red-shift is attributable to perturbations of the free-molecule potential surface in the adsorbed phase [11, 14]. Using a UV-sharpened fiber probe, DEZn was dissociated selectively and 20-nm Zn dots were fabricated successfully with 65 nm separation on a sapphire (0001) substrate (see Figure 3.4) [8]. Furthermore, since the nonresonant propagating light that leaked from the probe did not dissociate the gas-phase DEZn, atomic-level sapphire steps (see the line indicated by arrows A–A' in Figure 3.4) around the deposited dots were clearly observed after deposition.

To realize sub-10-nm scale controllability in size, the precise growth mechanism of Zn dots with NFO-CVD was investigated [8]. The deposition rate was found to be maximal when the dot grew to a size equivalent to the probe apex diameter. This dependence is accounted for by the theoretically calculated dipole–dipole coupling with a Förster field. The theoretical support and experimental results

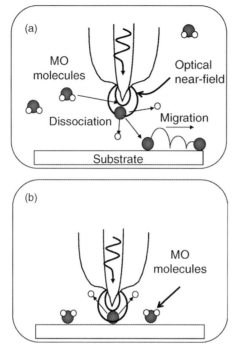

Figure 3.3 Schematic diagrams of the photodissociation of the (a) gas-phase and (b) adsorption-phase MO molecules.

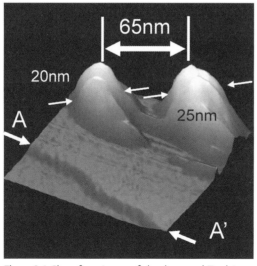

Figure 3.4 Shear-force image of closely spaced Zn dots.

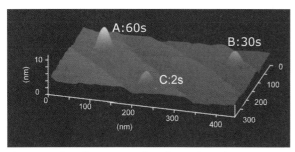

Figure 3.5 Shear-force image of Zn dots with irradiation times of 60 (dot A), 30 (dot B), and 5 s (dot C).

indicate the potential advantages of this technique for better regulating the size and position of deposited nanometer-scale dots.

As shown in Figure 3.2(d) the estimated apex diameter of the fiber probe used in this study, $2a_p$, was 9 nm based on the fitted dashed circle. Figure 3.5 shows a shear-force image of three Zn dots deposited with irradiation times of 60 (dot A), 30 (dot B), and 5 s (dot C) with a laser output power P of 5 µW.

Figure 3.6 plots the normalized deposition rate R of Zn dots as a function of the dot size S. Since the measured dot size S' was a convolution of the probe apex diameter $2a_p$ and the real size S, which was estimated as $S = S' - 2a_p$. Note that R is maximal at $S = 2a_p$. This indicates that the magnitude of the near-field optical interaction between the deposited Zn dot and the probe apex is enhanced resonantly with respect to S, resulting in the resonant increase in R. In other words, the near-field optical interaction exhibits size-dependent resonance characteristics.

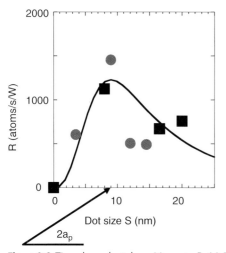

Figure 3.6 Time-dependent deposition rate R. (a) Experimental results. Solid squares and circles indicate the normalized deposition rate with a laser power P of 10 and 5 µW, respectively. Solid curves indicate the calculated values of I_2/I_1.

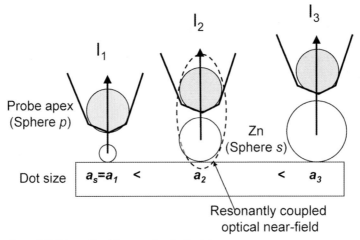

Figure 3.7 Schematic of the growth of a Zn dot.

To determine the origin of this size-dependent resonance, the magnitude of the near-field optical interaction was calculated between closely spaced nanoparticles (Figure 3.7). Spheres 'p' and 's' represent the probe apex and Zn dot, respectively. Since the separation between two particles is much narrower than the wavelength, the Förster field (proportional to R^{-3}, where R is the distance from the dipole) is dominant in the oscillating dipole electric field. In this quasistatic model, the intensity I_s of the light scattered from the two closely spaced spheres, 'p' and 's', is given by [15]:

$$I_s = I_1 + I_2 = (\alpha_p + \alpha_s)^2 |E|^2 + 4\delta\alpha(\alpha_p + \alpha_s)^2 |E|^2 \tag{3.1}$$

where $\alpha_i = 4\pi\varepsilon_0(\varepsilon_i - \varepsilon_0)/(\varepsilon_i + 2\varepsilon_0)a_i^3$ is the polarizability of sphere i ($=p, s$) with radius a_i. I_1 represents the light intensity scattered from the spheres and I_2 represents the light due to the dipole–dipole interaction induced by the Förster field. Therefore, the light intensity under study, normalized to I_1, is given by

$$I_2/I_1 = \frac{G_p A_p^3}{(A_p + 1)^3 (G_p A_p^3 + 1)} \tag{3.2}$$

where $A_p = a_p/a_s$ and $G_p = (\varepsilon_p - 1)(\varepsilon_s + 2)/(\varepsilon_p + 2)(\varepsilon_s - 1)$. For deposition by the fiber probe, the dielectric constants of Zn and fiber probe are $\varepsilon_s = (0.6 + i4)^2$ [16] and $\varepsilon_p = 1.52$, respectively. The diameter $2a_p$ of sphere p was 9 nm. The solid curve in Figure 3.6 show the calculated value of I_2/I_1 as a function of the Zn dot size S ($=2a_s$), which agrees well with the experimental results. This agreement indicates that the increase in R originates from the dipole–dipole coupling with the Förster field at a dot size equivalent to the probe apex diameter.

The experimental results and suggested mechanisms demonstrate the potential advantages of this technique for improving regulation of the size and position of deposited nanometer-scale dots.

3.3
Self-Assembling Method Via Optical Near-Field Interactions

This section demonstrates the size- and position-controlled deposition of nanometric materials based on the size-dependent resonance between the optical near-field and materials. This method of deposition enables highly precision nanofabrication without using an optical fiber probe and photomask.

3.3.1
Regulating the Size and Position of Nanoparticles Using Size-Dependent Resonance

To improve the size controllability, the dependence of plasmon resonance on the photon energy of optical near-fields can be used and the growth of nanoparticles can be controlled during the deposition process. Using this dependence, this section demonstrates the deposition of a nanometer-scale Zn dot using NFO-CVD [7].

First, nanoparticles were deposited on the cleaved facets of UV fibers (core diameter $= 10\,\mu$m) using conventional optical CVD. Gas-phase DEZn at a partial pressure of 5 mTorr was used as the source gas. The total pressure, including that of the Ar buffer gas, was 3 Torr. As the light source for the photodissociation of DEZn, a 500-μW He–Cd laser (photon energy $E_{p1} = 3.81$ eV [$\lambda = 325$ nm]) was coupled to the other end of the fiber. The irradiation time was 20 s. This irradiation covered the facet of the fiber core with a layer of Zn nanodots (see Figures 3.8(a) and 3.8(b)). Figure 3.9(a) shows a SEM image of the deposited Zn nanodots and their size distribution. The peak radius is 55 nm.

To control the size distribution, 20 μW Ar^+ ($E_{p2} = 2.54$ eV [$\lambda = 488$ nm]) or He–Ne ($E_{p2} = 1.96$ eV [$\lambda = 633$ nm]) laser light was introduced into the fiber, in addition to the He–Cd laser. Their photon energies are lower than the absorption band-edge energy of DEZn; that is, they are nonresonant light sources for the dissociation of DEZn. The irradiation time was 20 s. Figures 3.8(c) and (d), and Figures 3.8(e) and (f) show SEM images of Zn nanodots deposited with irradiation at $E_p = 3.81$ and 2.54 eV and at $E_p = 3.81$ and 1.96 eV, respectively. Figures 3.9(b) and (c) show the respective size distributions. The peak radii are 15 and 9 nm, respectively, which are smaller than those of the dots in Figure 3.9(a), and depend on the photon energy of the additional light. Furthermore, the FWHM was definitely narrower than that in Figure 3.9(a). These results suggest that the additional light controls the size of the dots and reduces the size fluctuation; that is, size regulation is realized.

Possible mechanisms for the size regulation of the dots using additional light are now discussed. A metal nanoparticle has strong optical absorption due to plasmon resonance [17, 18], which strongly depends on particle size. This can induce the desorption of the deposited metal nanoparticles [19, 20]. As the deposition of metal nanoparticles proceeds in the presence of light, the growth of the particles is affected by a trade-off between deposition and desorption, which determines their size, and depends on the photon energy. It has been reported that surface plasmon resonance in a metal nanoparticle is red-shifted with increasing particle size [19, 20]. However, the experimental results do not agree with these reports (compare Figures 3.9(a)–(c)).

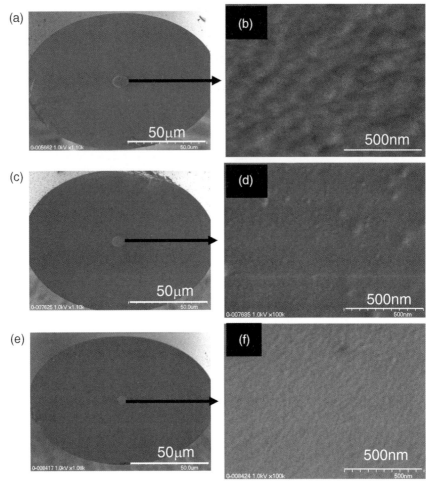

Figure 3.8 Conventional optical CVD on the cleaved facet of an optical fiber. (a) Schematic explanation. (b) SEM image of the end of the fiber. (a) and (b) $E_p = 3.81$ eV, (c) and (d) $E_p = 3.81$ and 2.54 eV, and (e) and (f) $E_p = 3.81$ and 1.96 eV.

To find the origin of this discrepancy, a series of calculations was performed and resonant sizes were evaluated. Mie's theory of scattering by a Zn sphere was used while considering the first mode only: [21]

$$\alpha = \frac{1 - \frac{1}{10}(\varepsilon + \varepsilon_m)x^2 + O(x^4)}{\left(\frac{1}{3} + \frac{\varepsilon_m}{\varepsilon - \varepsilon_m}\right) - \frac{1}{30}(\varepsilon + 10\varepsilon_m)x^2 - i\frac{4\pi^2 \varepsilon_m^{3/2}}{3}\frac{V}{\lambda_0^3} + O(x^4)} V \quad (3.3)$$

where $x(=\pi a_{\text{sphere}}/\lambda_0)$ is the size parameter, with a_{sphere} the diameter and λ_0 the wavelength in vacuum. The curves in Figure 3.10(a) represent the calculated polarizability α

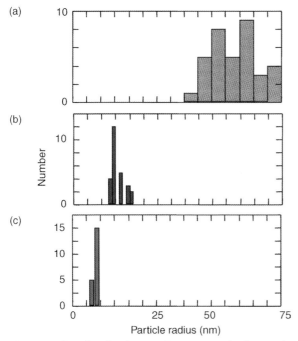

Figure 3.9 The radius distributions of Zn nanoparticles deposited using optical CVD with (a) $E_p = 3.81$ eV, (b) $E_p = 3.81$ and 2.54 eV, and (c) $E_p = 3.81$ and 1.96 eV.

with respect to three photon energies. The vertical axis is the value of α normalized to the volume, V, of a Zn sphere in air, which depends on its radius and is maximal at the resonant radius. The dashed curve in Figure 3.10(b) represents the resonant radius as a function of the photon energy, which is not a monotonous function and is minimal at $E_p = 2.0$ eV ($\lambda = 620$ nm). Since the imaginary part of the refractive index of Zn is also minimal at $E_p = 2.0$ eV ($\lambda = 620$ nm) (see the solid curve in Figure 3.10(b)), the minimum of the solid curve is due to the strong absorption in Zn.

Although Figure 3.10(a) shows that the resonant radius (47.5 nm) for $E_p = 2.54$ eV exceeds that (40 nm) for $E_p = 3.81$ eV, the calculated resonant radius for $E_p = 3.81$ eV is in good agreement with the experimentally confirmed particle size of 55 nm (see solid dot A in Figure 3.10(b)). Since the He–Cd laser light ($E_p = 3.81$ eV) is resonant for the dissociation of DEZn and is responsible for the deposition, irradiation with a He–Cd laser during deposition causes the particles to grow, and this growth halts when the particles reach the resonant radius because the rate of desorption increases due to resonant plasmon excitation. This is further supported by the fact that the resonant radius (37.5 nm) for $E_p = 1.96$ eV is smaller than that for $E_p = 3.81$ eV (see Figure 3.10(a)) and illumination with the additional light causes the particles to shrink (see Figures 3.9(b) and (c), and solid dots B and C in Figure 3.10(b)).

Another possible mechanism involves the acceleration of dissociation by the additional light. The photodissociation of DEZn produces transient monoethylzinc

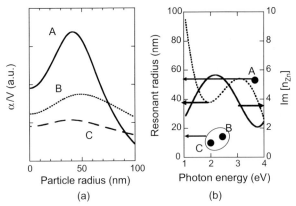

Figure 3.10 Calculated results. (a) Curves A–C show the polarizability α normalized to the volume V for a Zn sphere surrounded by air for $E_p = 3.81$, 2.51, and 1.96 eV, respectively. (b) The resonant radius of a Zn sphere (dashed curve). The imaginary part of the refractive index of Zn, n_{Zn}, used for the calculation (solid curve). Solid dots show the experimentally confirmed particle size. A: 3.81 eV, B: 2.54 eV, and C: 1.96 eV.

and Zn results from the dissociation of the monoethylzinc. Although the absorption band of monoethylzinc was not determined, the photon-energy dependence of the size observed using the additional light might have been due to the acceleration of the dissociation rate; that is, the additional light, which was nonresonant for DEZn, resonated the monoethylzinc [11], since the first metal–alkyl bond dissociation had a larger dissociation energy than the subsequent metal–alkyl bond dissociation [22, 23].

Based on the dependence described above, NFO-CVD was used to control the position of the deposited particle. Figures 3.11(a)–(c) show topographical images of Zn deposited by NFO-CVD with illumination from a 1-µW He–Cd laser ($E_p = 3.81$ eV) alone, or together with a 1-µW Ar$^+$ laser ($E_p = 2.54$ eV) or a 1-µW He–Ne laser ($E_p = 1.96$ eV), respectively. The irradiation times were 60 s. During deposition, the partial pressure of DEZn and the total pressure including the Ar buffer gas were maintained at 100 mTorr and 3 Torr, respectively. The respective FWHM was 60, 30, and 15 nm; that is, a lower photon energy gave rise to smaller particles, which is consistent with the experimental results shown in Figure 3.9.

These results suggest that the additional light controls the size of the dots and reduces the size fluctuation. Furthermore, the position can be controlled accurately by regulating the position of the fiber probe used to generate the optical near-field. The experimental results and suggested mechanisms described above show the potential advantages of this technique for controlling the size and position of the deposited nanodots. Furthermore, since our deposition method is based on a photodissociation reaction, it could be widely used for the nanofabrication of other materials, such as GaN [24, 25] and GaAs.

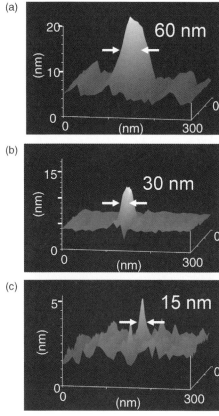

Figure 3.11 Bird's-eye view of shear-force topographical images of Zn deposited by NFO-CVD with (a) $E_p = 3.81$ eV, (b) $E_p = 3.81$ and 2.54 eV, and (c) $E_p = 3.81$ and 1.96 eV, respectively.

3.3.2
Self-Assembly of Nanoparticles Using Near-Field Desorption

To realize the mass production of nanometric structures, near-field desorption can be applied to other deposition techniques without using a fiber probe. An example is a self-assembling method that fabricates nanodot chains by controlling the desorption with an optical near-field [26]. This approach is illustrated schematically in Figure 3.12(a). A chain of metallic nanoparticles was fabricated using radio frequency (RF) sputtering under illumination on a glass substrate. To realize self-assembly, a simple groove 100 nm wide and 30 nm deep was fabricated on the glass substrate. During deposition of the metal, linearly polarized light illuminating the groove directly above (E_{90}) was used to excite a strong optical near-field at the edge of the groove (see Figure 3.12(b)), which induced the desorption of the deposited metallic nanoparticles [7]. A metallic dot has strong optical absorption due to plasmon resonance [17, 18], which strongly depends on the particle size. This can induce

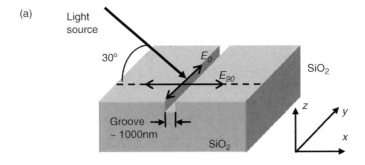
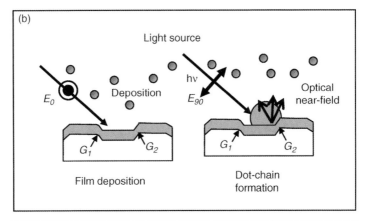
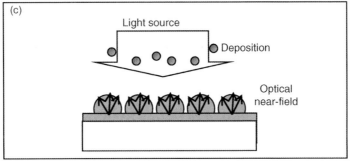

Figure 3.12 Size- and position-controlled formation of an ultralong nanodot chain. (a) The groove is parallel to the y-axis. The slanted light had a spot diameter of 1 mm. E_{90} and E_0 are perpendicular and parallel to the y-axis, respectively. (b), (c) Cross-sections in the x–z- and y–z-planes, respectively.

desorption of the deposited metallic nanodot when it reaches the resonant diameter [19, 20]. As the deposition of metallic dots proceeds, the growth is governed by a trade-off between deposition and desorption, which determines dot size, depending on the photon energy of the incident light. Consequently, the metallic nanoparticles should align along the groove (Figures 3.12(b) and (c)).

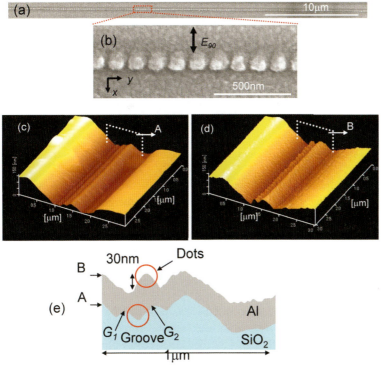

Figure 3.13 Experimental results. (a) SEM image of deposited Al with perpendicular polarization E_{90} ($h\upsilon = 2.33$ eV). (b) Magnified image of (a). (c) and (d) Respective AFM images of the surface of the glass substrate before and after aluminum deposition, at the same position. (e) Curves A and B show the respective cross-sectional profiles of AFM images across the groove before and after Al deposition, at the same position.

Illumination with 2.33-eV light (50 mW) during the deposition of aluminum (Al) resulted in the formation of 99.6-nm diameter Al nanodot chains with 27.9 nm separation that were as long as 100 μm in a highly size- and position-controlled manner (Figures 3.13(a) and (b)). The deviation of both nanodot size and the separation, determined from SEM images, was as little as 5 nm. To identify the position of the chain, we compared topographic atomic force microscopic (AFM) images of the surface of the glass substrate at the same position before (Figure 3.13(c)) and after (Figure 3.13(d)) Al deposition. Curves A and B in Figure 3.13(d) show the respective cross-sectional profiles across the groove. Comparison of these profiles showed that the nanodot chain formed around edge G_2. Furthermore, illumination with parallel polarization E_0 along the groove resulted in film growth along the groove structure and no dot structure was obtained. Since the near-field intensity with E_{90} was strongly enhanced at the metallic edge of the groove in comparison with E_0 owing to edge enhancement of the electrical field (see Figure 3.12(b)), a strong near-field intensity resulted in nanodot chain formation. Dot formation at the one-sided edge originated

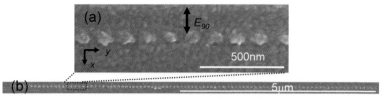

Figure 3.14 (a) SEM image of deposited Al with perpendicular polarization E_{90} ($h\upsilon = 2.62$ eV). (b) Magnified image of (a).

from the asymmetric electric-field intensity distribution, owing to the slanted illumination.

Chains of Al dots were also observed with RF sputtering of Al under illumination from 2.62-eV light (100 mW) with E_{90} using the same grooved (100 nm wide and 30 nm deep) glass substrate, which resulted in the formation of 84.2-nm nanodots with 48.6 nm separation (Figures 3.14(a) and (b)). Although the deviation of both nanodot size and the separation were as large as 10 nm, the dot size was reduced in proportion to the increase in the photon energy (99.6 nm × (2.33/2.62) = 88.5 − 84.2 nm). This indicates that the obtained size is determined by the photon energy and that the size-controlled dot-chain formation originates from photodesorption of the deposited metallic nanoparticles [7]. The period under 2.62-eV light illumination (132.8 nm) was longer than that (127.5 nm) using the 2.33-eV light. However, the ratios of the center–center distance (d) and radius (a) of the nanodots ($d/a = 2.56$ and 3.15 obtained under 2.33-eV and 2.62-eV light illumination, respectively) are similar to the optimum value. This is in the range of 2.4 to 3.0 for the efficient transmission of the optical energy along a chain of spherical metal dots calculated using Mie's theory [27]. This is determined by the trade-off between the increase in the transmission loss in the metal and the reduction in the coupling loss between adjacent metallic nanoparticles as the separation increases. To explain the optimum separation of the nanoparticles depending on the photon energy, theoretical analysis that includes the effect of the metallic film underneath the nanodot chain is required. However, these results imply that the center–center distance is set at the optimum distance for efficient energy transfer of the optical near-field, given that such a strong optical near-field can induce desorption of the deposited metallic nanoparticles and result in position-controlled dot-chain formation.

We anticipate that the fabricated structure will have high efficiency for optical near-field energy transfer, making it suitable as a nanodot coupler. Such efficient energy transfer has been reported along a nanodot chain with a metallic film underneath the nanodot chain [28, 29]. Furthermore, since our deposition method is based on a photodesorption reaction, illumination using a simple lithographically patterned substrate could realize the fabrication of size- and position-controlled nanoscale structures with other metals, e.g., Au (see Figure 3.15(a)) and Pt (see Figure 3.15(b)) or semiconductors. The use of the self-assembling method with a simple lithographically patterned substrate will dramatically increase the throughput of the production of nanoscale structures required by future systems.

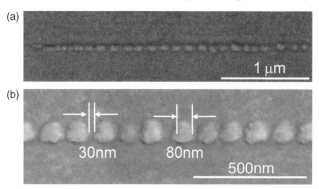

Figure 3.15 SEM images of (a) an Au dot-chain with 1.96-eV light illumination and (b) a Pt dot-chain with 2.33-eV light illumination.

3.3.3
One-Dimensional Alignment of Nanoparticles Using an Optical Near-field

Promising components for integrating the nanometer-sized photonic devices include chemically synthesized nanocrystals, such as metallic nanocrystals [30], semiconductor quantum dots [31], and nanorods [32], because they have uniform size, controlled shape, defined chemical composition, and tunable surface chemical functionality.

However, position- and size-controlled deposition methods have not yet been developed. Since several methods have been developed to prepare nanometer-sized templates reproducibly [33], it is expected that the self-assembly of colloidal nanostructures into a lithographically patterned substrate will enable precise control at all scales [34]. Capillary forces play an important role, because colloidal nanostructures are synthesized in solution. Recently, successful integration of polymer or silica spheres [34, 35] and complex nanostructures such as nanotetrapods [36] into templates by controlling the capillary force using appropriate template structures has been reported, although their size and separation are typically uniform. To fabricate nanophotonic devices, we propose a novel method of assembling nanoparticles by controlling the capillary force interaction and suspension flow. Further control of the positioning and separation of the nanoparticles is realized by controlling the particle–particle and particle–substrate interactions using an optical near-field.

To control position and separation very accurately, a preliminary experiment was performed on a patterned Si substrate, where an array of 10-μm holes in 100-nm thick SiO_2 was fabricated using photolithography (Figure 3.16(a)). Subsequently, a suspension containing latex beads with a mean diameter of 40 nm was dispersed on the substrate and the latex beads were aligned after solvent evaporation. The deposited latex beads were not subjected to any surface treatment and were dispersed in pure water at 0.001 wt%. Although the 10-μm sized template resulted in low selectivity in the position of the latex beads (Figures 3.16(b)–(d)), the beads were deposited only on the SiO_2 surface owing to its higher capillarity.

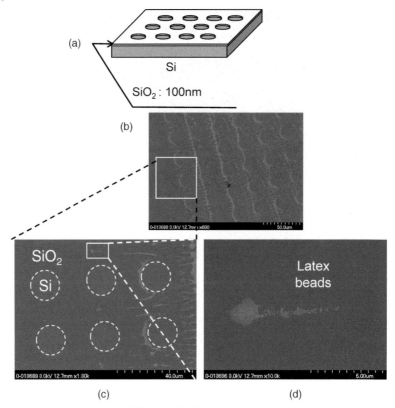

Figure 3.16 (a) Schematic of lithographically patterned Si substrate. (b)–(d) SEM images of latex beads dispersed on the lithographically patterned Si substrate

For higher positional selectivity, the suspension containing latex beads was dropped onto a lithographically patterned Si substrate that was spinning at 3000 revolutions per minute (rpm). As shown in Figure 3.17(a), the suspension flow split into two branches at the SiO$_2$ hole. SEM images (Figures 3.17 (a)–(d)) show that the chain of colloidal beads was aligned at the Si/SiO$_2$ interface. Note that the number of rows of latex beads decreased (Figures 3.17(c) and (d)) and only the smallest beads, which were 20 nm in diameter, reached the end of the suspension flow (Figure 3.17(d)).

Assuming the same particle-suspension contact angle (denoted ψ in Figure 3.17(e)) for various particle diameters, the flow speed of the larger latex beads had greater deceleration since the magnitude of the force pushing the particles on the SiO$_2$ (denoted F in Figure 3.17(e)) owing to evaporation of the solvent is proportional to the particle diameter [34]. In other words, the size selection was realized.

Based on the results of preliminary deposition, we tried assembling metallic nanoparticles because they are the material used to construct nanodot couplers [37].

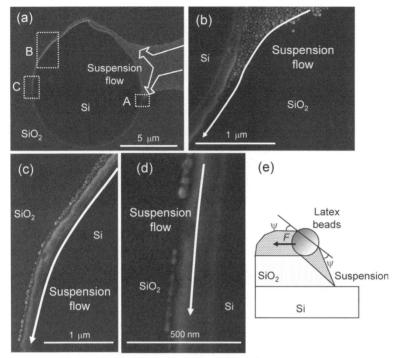

Figure 3.17 (a) SEM image of latex beads dispersed on the lithographically patterned Si substrate rotated at 3000 rpm. Higher-magnification SEM images of white squares A (b), B (c), and C (d) in (a). (e) Schematic illustrating of the particle-assembly process driven by the capillary force and suspension flow.

In this trial, we investigated the assembly of colloidal gold nanoparticles with a mean diameter of 20 nm dispersed in citrate solution at 0.001%. The nanoparticles were prepared by the citric acid reduction of gold ions and terminated by a carboxyl group (approximate length is 0.2 nm) with a negative charge [38]. However, they could not be aggregated using the same deposition process as for the latex beads (Figure 3.18(a)). To aggregate these particles, we fabricated a SiO_2 line structure with a plateau width of 50 nm on the Si substrate using photolithography. The solvent containing the colloidal gold nanoparticles was dropped onto this substrate at 3000 rpm. Then, the colloidal gold nanoparticles aggregated along the plateau of the SiO_2 line (Figures 3.18(b) and (c)).

This indicates that the capillary force induced by the lithographically patterned substrate, which is caused by the higher wettability of SiO_2 than that of the Si, was larger than the repulsive force owing to the negative charge of the carboxyl group on the colloidal gold nanoparticles, and this resulted in the aggregation and alignment of the colloidal gold nanoparticles at high density.

To further control size, separation, and positioning, we examined the aggregation of colloidal gold nanoparticles under illumination, because the colloidal gold

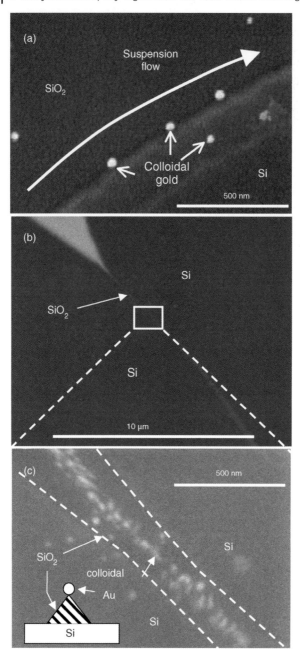

Figure 3.18 (a) SEM image of colloidal gold nanoparticles dispersed on the lithographically patterned Si substrate rotated at 3000 rpm. (b) and (c) SEM images of colloidal gold nanoparticles dispersed on the SiO_2 line rotated at 3000 rpm. Inset; cross-section of the substrate along the white line (dashes and dots).

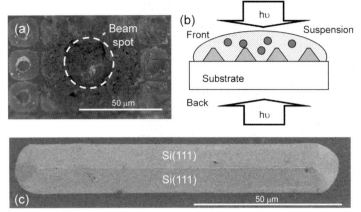

Figure 3.19 (a) Aggregated colloidal gold nanoparticles with frontal illumination under 690-nm light (25 mW/mm^2) for 60 s. (b) Schematic of the experimental setup. (c) SEM image of the fabricated Si wedge structure.

nanoparticles have strong optical absorption. Strong absorption should desorb the carboxyl group from the colloidal gold nanoparticles and result in their aggregation. Such an aggregation of colloidal gold nanoparticles was confirmed by the illumination of light. Figure 3.19(a) shows the aggregated gold nanoparticles over the pyramidal Si substrate under the 690-nm light illumination for 60 s. However, since the light was illuminated through the droplet of the colloidal gold nanoparticles, aggregated colloidal gold nanoparticles were spread outside the beam spot.

In order to realize selective aggregation of the gold nanoparticles at the desired position, the suspension was illuminated from behind (Figure 3.19(b)). Furthermore, we used a Si wedge, because this is a suitable structure for a far-/near-field conversion device [39]. The Si wedge structure (Figure 3.19(c)) was fabricated by the photolithography and anisotropical etching of Si.

For this structure, colloidal gold nanoparticles were deposited around the edge after evaporating the suspension without illumination (Figures 3.20(a) and (b)). Such aggregation is due to its wedge structure. This is because the suspension at the edge is thinner than that on the Si(1 1 1) plane owing to its low capillarity, and this causes the convective transport of particles toward the edge [40]. Further selective alignment along the edge of the Si wedge was realized using rear illumination. Figures 3.20(c) and (d) show the deposited colloidal gold nanoparticles with illumination under 690-nm light (25 mW/mm^2) for 60 s. Since the optical near-field energy is enhanced at the edge owing to the high refractive index of Si [41], selective aggregation along the edge with higher density is seen in these figures. This is due to the desorption of the carboxyl group by the absorption of light by the colloidal gold nanoparticles.

Note that the colloidal gold nanoparticles were closely aggregated and aligned linearly to form a wire shape when the polarization was perpendicular to the edge axis

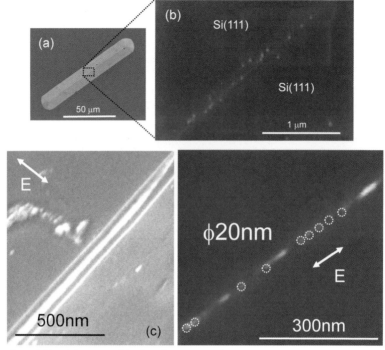

Figure 3.20 (a) Overview of the Si wedge structure. (b) SEM image of colloidal gold nanoparticles deposited on the edge of the Si wedge structure without illumination. SEM images of colloidal gold nanoparticles on the Si wedge structure under illumination with polarization (c) perpendicular and (d) parallel to the edge.

(Figure 3.20(c)), while they were aligned with a separation of several tens of nanometers in the parallel polarization (Figure 3.20(d)).

As the optical near-field energy for parallel polarization is higher than that for perpendicular polarization [39], greater aggregation is expected for parallel polarization. Nevertheless, the parallel polarization resulted in less aggregation. The low resolution of SEM images does not determine the distribution of the carboxyl molecules. However, such a repulsive force for disaggregation is caused by the carboxyl molecules that remained on the colloidal gold nanoparticles. Thus, we believe that the difference in the degree of aggregation originated from differences in the charge distribution induced inside the gold nanoparticles. Based on the polarization dependence of the aggregation, it is reasonable to consider that the aggregation along the edge with perpendicular polarization is owing to partially adsorbed carboxyl groups (Figure 3.21(a)), while the disaggregation with the parallel polarization resulted from the repulsive force induced by the partially attached carboxyl group on the colloidal gold nanoparticles (Figure 3.21(b)).

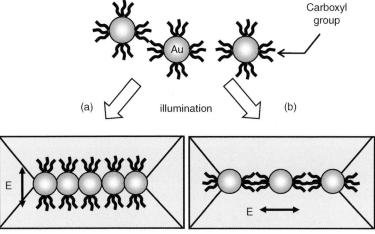

Figure 3.21 Schematic diagrams of the aggregation of colloidal gold nanoparticles along the edge of the Si wedge with polarization (a) perpendicular and (b) parallel to the edge.

3.4
Near-Field Imprint Lithography

As next-generation lithography (NGL) for the 32-nm node and below, ArF immersion lithography, extreme ultraviolet lithography (EUVL), and electron projection lithography (EPL) have been studied. However, a practical problem is the increasing cost and size of NGL tools. To solve these problems, optical near-field lithography has been developed by introducing trilayer resists. The fabrication of sub-50 nm features has been realized using the i-line ($\lambda = 365$ nm) [42–44], with conventional photolithography facilities.

A further decrease in feature size has been reported with the introduction of imprint lithography, which resulted in the fabrication of 14-nm pitch features [33]. Although conventional imprint lithography results in the same mold structure, the use of the optical near-field intensity distribution should realize features smaller than the mold structure, since a nanoscale mold has a nanoscale optical field distribution at its edge.

In this chapter, we propose and demonstrate optical near-field imprint lithography to introduce its ability to obtain a higher resolution than the size of the mold features [45].

To realize the efficient excitation of an optical near-field on a mold, the thickness of the metallic film and the coverage were optimized using the finite-difference time-domain (FDTD) method [46]. For comparison with a conventional photolithographic mask, the optical field distribution for 80-nm half-pitch, 200-nm deep Cr line-and-space (LS) on the SiO_2 ($n = 1.5$) substrate was calculated at a wavelength of 436 nm (g-line). Here, the refractive index of Cr was assumed to be $n = 1.78 + i2.69$ [47], and the line was parallel to the y-axis (Figure 3.22(a)). Since the imprint mold

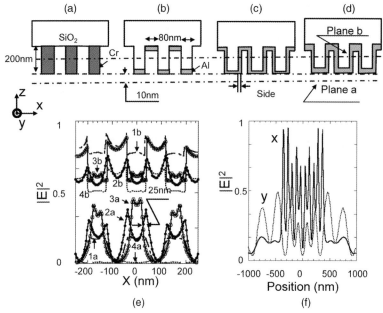

Figure 3.22 Schematic of the calculation models. (a) Cr LS on SiO$_2$ substrate. (b)–(d) SiO$_2$ LS coated with (b) 20-nm Al without a sidewall coating, (c) 20-nm Al with a sidewall coating, and (d) 50-nm Al at the top and 20-nm Al on the sidewall. (e) Curves 1a–4a show the cross-sectional profiles along the x-axis 10 nm from the mold (plane a) with five grooves in the mold. Curves 1b–4b show the cross-sectional profiles along the x-axis 10 nm from the bottom of the mold (plane b). The beam width at 1/e^2 of incident light with a Gaussian shape was 1000 nm. The x-coordinate is perpendicular to the grating in (a)–(d). (f) The polarization dependence of the cross-sectional profiles for x- (perpendicular to the LS direction) and y- (parallel to the LS direction) polarization for the mold with five grooves in the mold. The beam width at 1/e^2 of incident light with a Gaussian shape was 2000 nm (1/e^2).

used in this study was fabricated using SiO$_2$, we also calculated the optical-field distribution of 80-nm half-pitch, 200-nm deep SiO$_2$ LS, coated with aluminum film ($n = 0.56 + i5.2$ [47]) (Figures 3.22(b)–(d)). The minimum cell size was $5 \times 12.5 \times 5$ nm.

Figure 3.22(e) shows the cross-sectional profile along the x-axis 10 nm from the mold. The optical field intensity distribution of the Cr LS used for conventional photolithography resulted in a single peak corresponding to the space of the Cr mask, which resulted from reducing the optical field intensity through the 200-nm thick Cr film (curve 1a in Figure 3.22(e)). By contrast, coating the SiO$_2$ LS with a 20-nm thick Al film (Figures 3.22(b) and (c)) enhanced the electric field intensity at the edge of the mold (curves 2a and 3a in Figure 3.22(e)). Higher localization at the edge of the mold was realized without a sidewall coating (Figure 3.22(b) and curve 2a in Figure 3.22(e)). Since this localization was not observed for the thicker coating (a 50-nm thick Al film on top of SiO$_2$ and a 20-nm thick Al film on the sidewall) and was not observed in the y-polarization (curve y in Figure 3.22(f)), this localized optical near-field originated from the edge effect. Since efficient excitation of the surface plasmon is obtained with a 15-nm thick aluminum coating on glass in the Kretschmann configuration [48], this size-dependent feature is attributed to localized surface plasmon resonance on the Al film. Furthermore, the localization of the optical field intensity to an area as narrow as 25 nm in curve 2a in Figure 3.22(e) infers the realization of a resolution higher than the mold size. This localization was also observed in curve 2b (Figure 3.22(e)) obtained in the plane along the x-axis 10 nm from the bottom of the mold.

We performed imprint lithography to confirm the higher-resolution capability using an optical near-field, as discussed above. Commercial photocurable acryl PAK01 resin (blended by Toyo Gosei) was used; it is composed of tri-propylene-glycol-diacrylate monomer with dimethoxy-phenyl-acetophenon as the photoinitiator and has good release properties [49]. Polycarbonate (PC) substrate was spin-coated with PAK01. We used 300-nm half-pitch, 200-nm deep SiO_2 LS as the mold.

To obtain the optimum structure shown in Figure 3.1(b) (20-nm thick Al film with no sidewall coating), the mold was coated with Al using vacuum evaporation (Figure 3.23(a)). The mold was pressed into the liquid polymer on PC substrate

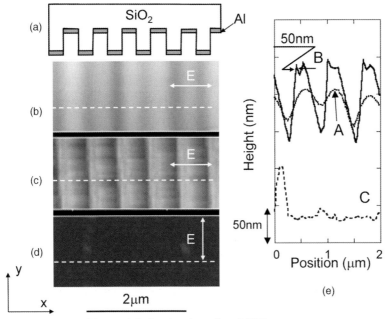

Figure 3.23 (a) Mold position before its release from PAK01. (b), (c), and (d) show AFM images of the surface of the PAK01 using a bare mold with x-polarization (perpendicular to the LS direction), an Al-coated mold with x-polarization, and an Al-coated mold with y-polarization (parallel to the LS direction), respectively. (d) Curves A, B, and C show the cross-sectional profiles along the dashed white lines in (b), (c), and (d), respectively.

under a pressure of 70 kPa using a conventional contact mask aligner (MJB3, SUSS MicroTec KK). It was irradiated with UV light (g-line: 436 nm, power density: 30 mW/cm^2) for 30 s from the back of the mold, while maintaining the imprint pressure during exposure. After pressing the mold and UV curing, the PC substrate was separated from the mold, and the pattern was transferred.

First, we obtained topographic atomic force microscopy (AFM) images of the surface of PAK01 after release of the mold. Figures 3.23(b)–(d) show AFM images of a bare SiO$_2$ mold with x-polarization (perpendicular to the LS direction), an Al-coated SiO$_2$ mold with x-polarization, and an Al-coated SiO$_2$ mold with y-polarization (parallel to the LS direction), respectively. Although the bare mold resulted in a single pitch corresponding to the mold pitch (Figure 3.23(b) and curve A in Figure 3.23(e)), we obtained sharp (50 nm) protruding structures at the edge of the Al-coated SiO$_2$ mold, when the mold was pressed under x-polarization (Figure 3.23(c) and curve B in Figure 3.23(e)). Although the pitch differs between the numerical and experimental results, these profiles with protruding structures seen at the edge of the Al-coated SiO$_2$ mold are in good agreement with those calculated using FDTD (curve 2a of Figure 3.22(e)). The calculated value at the point next to the interface is unstable in the FDTD calculation due to the drastic change in the refractive index. Furthermore, since the distribution of the optical near-field along the z-axis is as large as that along the x-axis, the optical near-field is believed to be localized to a region as small as 20 nm along the z-axis. Therefore, we compared the profiles at the second point from the interface (10 nm apart from the interface). Although the calculated and obtained profiles differ, the protuberances were obtained only with x-polarization with an Al coating; therefore, we believe that the protuberances originated from the plasmon resonance, as predicted by the FDTD calculation. These results indicate that the resolution was higher than the pitch of the mold.

Next, we obtained scanning electron microscope (SEM) images of the transferred pattern (Figures 3.24(a)–(c)). As shown in the AFM images, the SEM images confirmed that there were protuberances where the edge of the mold was pressed (inside the white solid ellipses in Figure 3.24(b.1) and 3.24(b.2)) using the Al-coated mold with x-polarization.

The field distribution calculated using the FDTD method did not predict the resist structure after release of the mold. However, the correspondence between the AFM and SEM images showed that protuberances were formed at positions where a strong optical near-field was localized. Based on these results, we concluded that using conventional imprint lithography, a strong uniform optical field intensity resulted in the formation of the same pitch as that of the mold (Figure 3.25(a)). In contrast, with optical near-field imprint lithography, the resist was deformed in the underexposed condition where the flat surface of the mold was pressed, which was due to the Al coating of the mold, and the resist was remained in the overexposed condition arising from the strong optical near-field, which was due to the edge effect from where the edge of the mold was pressed. This enhancement originated from the resonant excitation of the surface plasmon on the Al film, which resulted in protuberances smaller than the pitch of the mold (see Figure 3.25(b)). Future evaluations, which will

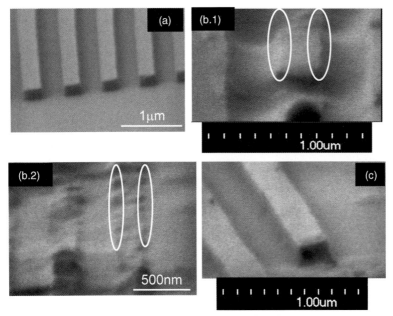

Figure 3.24 Tilted (30°) SEM images using (a) a bare mold with x-polarization (perpendicular to the LS direction), (b.1) (b.2) an Al-coated mold with x-polarization, and (c) an Al-coated mold with y-polarization.

include the effects of power and irradiation time, are required to explain the optimum dose for the higher contrast of the protuberances.

We performed optical near-field imprint lithography to increase the resolution over conventional imprint lithography. By introducing local field enhancement of the

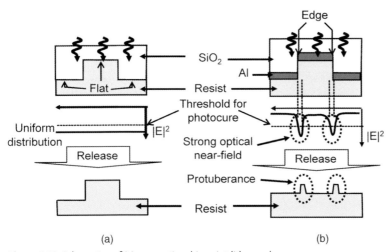

Figure 3.25 Schematics of (a) conventional imprint lithography and (b) optical near-field imprint lithography.

optical near-field using a metallized LS SiO$_2$ mold (300-nm half-pitch and 200 nm deep) without sidewall coating, we obtained features as small as 50 nm wide on illumination with the g-line ($\lambda = 436$ nm). The resolution could be increased further by surface treatment with Al or other durable metals or semiconductors.

3.5
Nonadiabatic Optical Near-Field Etching

An ultraflat surface substrate of sub-nm scale roughness is required for various applications including the manufacture of high-quality, extreme UV optical components, high-power lasers, ultrashort pulse lasers, plus future photonic devices at the sub-100 nm scale; it is estimated that the required surface roughness, R_a, will be less than 1 Å [50]. This R_a value is an arithmetic average of the absolute values of the surface height deviations measured from the best-fitting plane, and is given by

$$R_a = \frac{1}{l} \int_0^l |f(x)| dx$$
$$\cong \frac{1}{n} \sum_{i=1}^{n} |f(x_i)|$$
(3.4)

where the $|f(x_i)|$ are absolute values measured from the best-fitting plane and l is the evaluation length. Physically, dx corresponds to the spatial resolution in the measurement of $f(x)$ and n is the number of pixels in the measurement; $n = l/dx$. Conventionally, chemical-mechanical polishing (CMP) has been used to flatten the surface [51]. However, CMP has difficulties in reducing R_a to less than 2 Å because the polishing pad roughness is as large as 10 μm and the diameters of the polishing particles in the slurry are as large as 100 nm. In addition, polishing causes scratches or digs due to the contact between the polishing particles and/or impurities in the slurry and the substrate.

Our interest in applying an optical near-field to nanostructure fabrication was generated because of its high resolution capability, beyond the diffraction limit, and its novel photochemical reaction, which is classified as nonadiabatic due to its energy transfer via a virtual exciton-phonon-polariton [52, 53]. In this chemical vapor deposition, photodissociation of the molecules is driven by the light source at a lower photon energy than the molecular absorption band-edge energy by a multiple step excitation via vibrational energy levels [54]. Following this process, we proposed a novel method of polishing using nonadiabatic optical near-field etching [55].

A continuum-wave laser ($\lambda = 532$ nm) was used to dissociate the Cl$_2$ gas through a nonadiabatic photochemical reaction. This photon energy is lower than the absorption band-edge energy of Cl$_2$ ($\lambda = 400$ nm) [56], so that the conventional Cl$_2$ adiabatic photochemical reaction is avoided. However, because the substrate has nanometer-scale surface roughness, the generation of a strong optical near-field

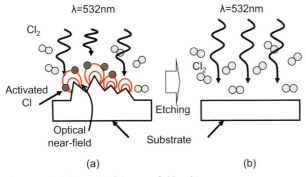

Figure 3.26 Schematic of the near-field etching.

on the surface is expected from the simple illumination with no focusing being required (Figure 3.26(a)). Since a virtual exciton-phonon-polariton can be excited on this roughness, a higher molecular vibrational state can be excited than on the flat part of the surface, where there are no virtual exciton-phonon-polaritons. Cl_2 is therefore selectively photodissociated wherever the optical near-field is generated. These dissociated Cl_2 molecules then etch away the surface roughness; the etching process automatically stops when the surface becomes flat (Figure 3.26(b)).

We used 30-mm diameter planar synthetic silica substrates built by vapor-phase axial deposition with an OH group concentration of less than 1 ppm [57]. The substrates were preliminarily polished by CMP prior to the nonadiabatic optical near-field etching. We performed the nonadiabatic optical near-field etching at a Cl_2 pressure of 100 Pa at room temperature with a continuum-wave laser ($\lambda = 532$ nm) having a uniform power density of 0.28 W/cm^2 over the substrate (see Figure 3.27). Surface roughness was evaluated using an atomic force microscope (AFM). Since the scanning area of the AFM was much smaller than the substrate, we measured the surface roughness R_a in nine representative areas, each 10 μm × 10 μm, separated

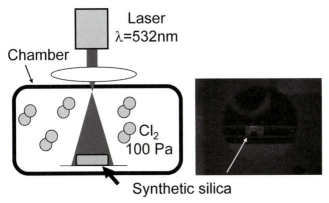

Figure 3.27 Schematic of the experimental setup.

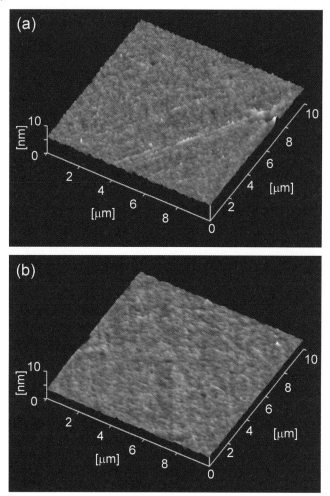

Figure 3.28 Typical AFM images of the silica substrate (a) before and (b) after nonadiabatic optical near-field etching.

by 100 μm. The scanned area was 256 × 256 pixels with a spatial resolution of 40 nm. The average value $\bar{R}_a$ of the nine R_as, obtained before the nonadiabatic optical near-field etching, and evaluated through the AFM images, was 2.36 ± 0.02 Å. We cleaned the substrate ultrasonically using deionized water and methanol before and after the nonadiabatic optical near-field etching.

Figures 3.28(a) and 3.28(b) show typical AFM images of the scanned 10 μm × 10 μm silica substrate area before and after nonadiabatic optical near-field etching, respectively. Note that the surface roughness was drastically decreased, as supported by the cross-sectional profiles along the dashed white lines in Figures 3.28(a) and 3.28(b) (see Figure 3.29). We found a dramatic decrease in the value of the peak-to-valley from 1.2 nm (curve B) to 0.5 nm (curve A). Furthermore, note that the scratch seen

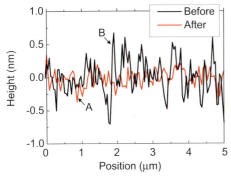

Figure 3.29 Cross-sectional profiles along the white dashed lines in (a) and (b). Curves A and B correspond to the profile after and before etching, respectively.

in the AFM image before nonadiabatic optical near-field etching disappeared. This indicates that rougher areas of the substrate had a higher etching rate, possibly because of the greater intensity of the optical near-field, leading to a uniformly flat surface over a wide area.

Figure 3.30 shows the etching time dependence of $\bar{R}_a$. We found that $\bar{R}_a$ decreases as the etching time increases. The minimum in $\bar{R}_a$ was 1.37 Å at an etching time of 120 min, while the minimum R_a among the nine areas was 1.17 Å. Because the process is performed in a sealed chamber, the saturation in the decrease of $\bar{R}_a$ might originate from the decrease in Cl_2 partial pressure during etching, A further decrease in $\bar{R}_a$ would be expected under constant Cl_2 pressure.

We propose a new polishing method that uses near-field etching based on a nonadiabatic process, with which we obtained an ultraflat silica surface that had a minimum roughness of 1.37 Å. We believe our technique is applicable to a variety of substrates, including amorphous and crystal substrate. Since this technique is a noncontact method without a polishing pad, it can be applied not only to flat

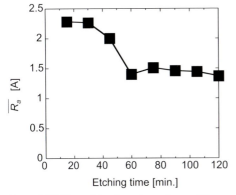

Figure 3.30 The etching-time dependence of the average $R_a(\bar{R}_a)$.

substrates but also to three-dimensional substrates that have convex or concave surfaces, such as microlenses and the inner-wall surface of cylinders. Furthermore, this method is also compatible with mass production.

References

1 For example, see the International Technology Roadmap for Semiconductors (http://public.itrs.net/).
2 Ohtsu, M., Kobayashi, K., Kawazoe, T., Sangu, S. and Yatsui, T. (2002) Nanophotonics: design,fabrication, and operation of nanometric devices using optical near-fields. *IEEE Journal of Selected Topics in Quantum Electronics*, **8**, 839–862.
3 Polonski, V.V., Yamamoto, Y., Kourogi, M., Fukuda, H. and Ohtsu, M. (1999) Nanometric patterning of zinc by optical near-field photochemical vapor deposition. *Journal of Microscopy*, **194**, 545–551.
4 Yamamoto, Y., Kourogi, M., Ohtsu, M., Polonski, V.V. and Lee, G.H. (2000) Fabrication of nanometric zinc pattern with photodissociated gas-phase diethylzinc by optical near-field. *Applied Physics Letters*, **76**, 2173–2175.
5 Yatsui, T., Kawazoe, T., Ueda, M., Yamamoto, Y., Kourogi, M. and Ohtsu, M. (2002) Fabrication of nanometric single zinc and zinc oxide dots by the selective photodissociation of adsorption-phase diethylzinc using a nonresonant optical field. *Applied Physics Letters*, **81**, 3651–3653.
6 Yamamoto, Y., Kourogi, M., Ohtsu, M., Lee, G.H. and Kawazoe, T. (2002) Lateral integration of Zn and Al dots with nanometer-scale precision by near-field optical chemical vapor deposition using a sharpened optical fiber probe. *IEICE Transactions on Electronics*, **E85-C**, 2081–2085.
7 Yatsui, T., Takubo, S., Lim, J., Nomura, W., Kourogi, M. and Ohtsu, M. (2003) Regulating the size and position of deposited Zn nanoparticles by optical near-field desorption using size-dependent resonance. *Applied Physics Letters*, **83**, 1716–1718.
8 Lim, J., Yatsui, T. and Ohtsu, M. (2005) Observation of size-dependent resonance of near-field coupling between a deposited Zn dot and the probe apex during near-field optical chemical vapor deposition. *IEICE Transactions on Electronics*, **E 88-C**, 1832–1835.
9 Mononobe, S. (1999) Probe fabrication, in *Near-Field Nano/Atom Optics and Technology*, (ed. M. Ohtsu), Springer-Verlag (Berlin), pp. 31–69.
10 Polonski, V.V., Yamamoto, Y., White, J.D., Kourogi, M. and Ohtsu, M. (1999) Vacuum shear force microscopy application to high resolution work. *Japanese Journal of Applied Physics*, **38**, L826–L829.
11 Krchnavek, R.R., Gilgen, H.H., Chen, J.C., Shaw, P.S., Licata, T.J. and Osgood, R.M. Jr (1987) Photodeposition rates of metal from metal alkyls. *Journal of Vacuum Science & Technology B*, **5**, 20–26.
12 Jackson, R.L. (1989) Metal-alkyl bond dissociation energies for $(CH_3)_2Zn$, $(C_2H_5)Zn$, $(CH_3)_2Cd$, and $(CH_3)_2Hg$. *Chemical Physics Letters*, **163**, 315–322.
13 Yatsui, T., Kawazoe, T., Shimizu, T., Yamamoto, Y., Ueda, M., Kourogi, M., Ohtsu, M. and Lee, G.H. (2002) Fabrication of nanometric single zinc and zinc oxide dots by the selective photodissociation of adsorption-phase diethylzinc using a nonresonant optical field. *Applied Physics Letters*, **80**, 1444–1446.
14 Chen, C.J. and Osgood, R.M. Jr (1983) Direct Observation of the Local-Field-Enhanced Surface Photochemical Reactions. *Physical Review Letters*, **50**, 1705–1708.

15 Ohtsu, M. and Kobayashi, K. (2003) *Optical Near-Fields*, Springer-Verlag (Berlin).
16 Yarovaya, R.G., Shklyarevsklii, I.N. and El-Shazly, A.F.A. (1974) Temperature dependence of the optical properties and the energy spectrum of zinc. *Soviet Physics. JETP*, **38**, 331–334.
17 Boyd, G.T., Rasing, T., Leite, J.R.R. and Shen, Y.R. (1984) Local-field enhancement on rough surfaces of metals, semimetals, and semiconductors with the use of optical second-harmonic generation. *Physical Review B-Condensed Matter*, **30**, 519–526.
18 Wokaum, A., Gordon, J.P. and Liao, P.F. (1982) Radiation damping in surface-enhanced Raman scattering. *Physical Review Letters*, **48**, 957–960.
19 MacDonald, K.F., Fedotov, V.A., Pochon, S., Ross, K.J., Stevens, G.C., Zheludev, N.I., Brocklesby, W.S. and Emel'yanov, V.I. (2002) Optical control of gallium nanoparticle growth. *Applied Physics Letters*, **80**, 1643–1645.
20 Bosbach, J., Martin, D., Stietz, F., Wenzel, T. and Träger, F. (1999) Laser-based method for fabricating monodisperse metallic nanoparticles. *Applied Physics Letters*, **74**, 2605–2607.
21 Kuwata, H., Tamaru, H. and Miyano, K. (2003) Resonant light scattering from metal nanoparticles: Practical analysis beyond Rayleigh approximation. *Applied Physics Letters*, **83**, 4625–4627.
22 Sato, A., Tanaka, Y., Tsunekawa, M., Kobayashi, M. and Sato, H. (1993) Laser photodissociation of Trimethylgallium in the gas phase and on a quartz substrate as studied by multiphoton ionization-time of flight mass spectroscopy. *The Journal of Physical Chemistry*, **97**, 8458–8463.
23 Young, P.J., Gosavi, R.K., Connor, J., Strausz, O.P. and Gunning, H.E. (1973) Ultraviolet absorption spectra of $CdCH_3$, $ZnCH_3$, and $TeCH_3$. *Journal of Chemical Physics*, **58**, 5280–5283.
24 Yamazaki, S., Yatsui, T., Ohtsu, M., Kim, T.W. and Fujioka, H. (2004) Room-temperature synthesis of ultraviolet-emitting nanocrystalline GaN films using photochemical vapor deposition. *Applied Physics Letters*, **85**, 3059–3061.
25 Yamazaki, S., Yatsui, T. and Ohtsu, M. (2008) Room-temperature growth of ultraviolet-emitting GaN with a hexagonal crystal-structure using photochemical vapor deposition. *Applied Physics Express*, **1**, 061102.
26 Yatsui, T., Nomura, W. and Ohtsu, M. (2005) Self-assembly of size- and position-controlled ultralong nanodot chains using near-field optical desorption. *Nano Letters*, **5**, 2548–2551.
27 Quinten, M., Leitner, A., Krenn, J.R. and Aussenegg, F.R. (1998) Electromagnetic energy transport via linear chains of silver nanoparticles. *Optics Letters*, **23**, 1331–1333.
28 Nomura, W., Yatsui, T. and Ohtsu, M. (2005) Nanodot coupler with a surface plasmon polariton condenser for optical far/near-field conversion. *Applied Physics Letters*, **86**, 181108.
29 Nomura, W., Yatsui, T. and Ohtsu, M. (2006) Efficient optical near-field energy transfer along an Au nanodot coupler with size-dependent resonance. *Applied Physics B-Lasers and Optics*, **84**, 257–259.
30 Brust, M. and Kiely, C.J. (2002) Some recent advances in nanostructure preparation from gold and silver particles: short topical review. *Colloids Surf A*, **202**, 175–186.
31 Alivisatos, A.P. (1996) Semiconductor clusters, nanocrystals, and quantum dots. *Science*, **271**, 933–937.
32 Yang, P. (2003) Wires on water. *Nature*, **425**, 243–244.
33 Austin, M.D., Ge, H., Wu, W., Li, M., Yu, Z., Wasserman, D., Lyon, S.A. and Chou, S.Y. (2004) Fabrication of 5 nm linewidth and 14 nm pitch features by nanoimprint lithography. *Applied Physics Letters*, **84**, 5299–5301.
34 Whitesides, G.M. and Grzybowski, B. (2002) Self-assembly at all scales. *Science*, **295**, 2418–2421.
35 Yin, Y., Lu, Y. and Xia, Y. (2001) Assembly of monodispersed spherical colloids into

one-dimensional aggregates characterized by well-controlled structures and length. *Journal of Materials Chemistry*, **11**, 987–989.
36 Cui, Y., Björk, M.T., Liddle, J.A., Sönnichsen, C., Boussert, B. and Alivisatos, A.P. (2004) Integration of colloidal nanocrystals into lithographically patterned devices. *Nano Letters*, **4**, 1093–1098.
37 Maier, S.A., Kik, P.G., Atwater, H.A., Meltzer, S., Harel, E., Koel, B.E. and Requicha, A.G. (2003) Local detection of electromagnetic energy transport below the diffraction limit in metal nanoparticle plasmon waveguides. *Nature Materials*, **2**, 229–232.
38 Frens, G. (1973) Controlled nucleation for the regulation of the particle size in monodisperse gold suspensions. *Nature Physical Science*, **241**, 20–22.
39 Yatsui, T., Kourogi, M. and Ohtsu, M. (2001) Plasmon waveguide for optical far/near-field conversion. *Applied Physics Letters*, **79**, 4583–4585.
40 Denkov, N.D., Velev, O.D., Kralchevsky, P.A., Ivanov, I.B., Yoshimura, H. and Nagayama, L. (1993) Two-dimensional crystallization. *Nature*, **361**, 26.
41 Yatsui, T., Itsumi, K., Kourogi, M. and Ohtsu, M. (2002) Metallized pyramidal silicon probe with extremely high throughput and resolution capability for optical near-field technology. *Applied Physics Letters*, **80**, 2257–5529.
42 Ito, T., Ogino, M., Yamada, T., Inao, Y., Yamaguchi, T., Mizutani, T. and Kuroda, R. (2003) Fabrication of sub-100 nm Patterns using Near-field Mask Lithography with Ultra-thin Resist Process. *Journal of Photopolymer Science and Technology*, **18**, 435–441.
43 Yonemitsu, H., Kawazoe, T., Kobayashi, K. and Ohtsu, M. (2007) Nonadiabatic photochemical reaction and application to photolithography. *Journal of Luminescence*, **122–123**, 230–233.
44 Kawazoe, T., Ohtsu, M., Inao, Y. and Kuroda, R. (2007) Exposure dependence of the developed depth in nonadiabatic photolithography using visible optical near-fields. *Journal of Nanophotonics*, **1**, 011595.
45 Yatsui, T., Nakajima, Y., Nomura, W. and Ohtsu, M. (2006) High-resolution capability of optical near-field imprint lithography. *Applied Physics B-Lasers and Optics*, **84**, 265–267.
46 The computer simulations in this paper are performed by a FDTD-based program, *Poynting for Optics*, a product of Fujitsu, Japan.
47 Palik, E.D. (ed.) (1985) *Handbook of Optical Constants of Solids*, Academic, New York.
48 Raether, H. (ed.) (1988) *Surface Plasmons*, Springer-Verlag, Berlin.
49 Haisma, J., Verheijen, M., van den Heuvel, K. and van den Berg, J. (1996) Mold-assisted nanolithography: A process for reliable pattern replication. *Journal of Vacuum Science & Technology B*, **14**, 4124–4128.
50 Wua, B. and Kumar, A. (2007) Extreme ultraviolet lithography: A review. *Journal of Vacuum Science & Technology B*, **25**, 1743.
51 Cook, L.M. (1990) Chemical process in glass polishing. *Journal of Non-Crystalline Solids*, **120**, 152.
52 Kawazoe, T., Kobayashi, K., Takubo, S. and Ohtsu, M. (2005) Nonadiabatic photodissociation process using an optical near-field. *Journal of Chemical Physics*, **122**, 024715.
53 Kobayashi, K., Kawazoe, T. and Ohtsu, M. (2005) Importance of multiple-phonon interactions in molecular dissociation and nanofabrication using optical near-fields. *IEEE Transactions on Nanotechnology*, **4**, 517–522.
54 Kawazoe, T., Yamamoto, Y. and Ohtsu, M. (2001) Fabrication of nanometric Zn dots by nonresonant near-field optical chemical-vapor deposition. *Applied Physics Letters*, **79**, 1184–1186.
55 Yatsui, T., Hirata, K., Nomura, W., Ohtsu, M. and Tabata, Y. (2008) Realization of an

ultraflat silica surface with angstrom-scale average roughness using nonadiabatic optical near-field etching. *Applied Physics B-Lasers and Optics*, in press.

56 Kullmer, R. and Büerle, D. (1987) Laser-induced chemical etching of silicon in chlorine atmosphere. *Applied Physics A-Materials Science & Processing*, **43**, 227–232.

57 Izawa, T. and Inagaki, N. (1980) Materials and processes for fiber preform fabrication-vapor phase axial deposition. *Proceedings of the IEEE*, **68**, 1184–1187.

4
Fabrication of Quantum Dots for Nanophotonic Devices
Kouichi Akahane and Naokatsu Yamamoto

4.1
Introduction

A semiconductor quantum dot (QD) is a three-dimensional nanoscale structure that confines electrons, holes, and excitons. Structures that can restrict the spatial degree of freedom of electrons and other particles are referred to as quantum structures. When the restricted spatial degrees of freedom are one-dimensional, two-dimensional, and three-dimensional, the structures are referred to as a quantum well (QW), quantum wire (QWr), and QD, respectively. As shown in Figure 4.1, the density of states of these quantum structures—in other words, the number of states at a certain energy—changes from a parabolic shape for the bulk structure to a corresponding step shape (QW), saw-tooth shape (QWr), or delta-function shape (QD). Accordingly, the behaviors of light absorption and emission are considered to change sequentially, which is expected to induce a change in the optical response.

In QDs, the carriers are concentrated at a certain energy value, theoretically yielding laser diodes with extremely low thresholds. Moreover, the temperature dependence of the threshold current disappears when using QDs in semiconductor lasers: this characteristic can be attributed to the change in the density of states of QDs. In other words, the density of states changes continuously in a bulk material or in a QW structure. (The density of states of a QW changes stepwise at certain energy values but maintains a constant value at other energy values.) Therefore, when the device temperature increases, the injected carriers are redistributed, and the carrier density—which contributes to laser oscillation—decreases. As a result, new carriers need to be injected for laser oscillations. Therefore, the threshold current required for laser oscillations generally shows a tendency to increase as the temperature increases. On the other hand, a QD laser features a delta-function-form density of states; therefore, only discrete values are allowed to represent the carrier energy, even if the carriers attempt to redistribute themselves when the temperature increases. In other words, the carrier redistribution is suppressed. This suppression maintains the density of carriers of a certain energy value before and after the

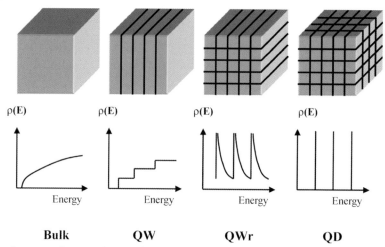

Figure 4.1 Change in density of states associated with decrease in structure dimensionality.

temperature increases, which suppresses the increase in the threshold current that would otherwise result from the increase in temperature. Therefore, we can create a situation in an ideal QD laser in which the threshold current is completely independent of temperature [1]. Of course, it is difficult to fabricate completely ideal QDs, and we have not yet produced a laser with a threshold current that is completely independent of temperature. Nevertheless, among the semiconductor lasers available today, a QD laser has the least temperature dependence on the threshold current. When we use QDs, we can implement a high-performance semiconductor laser that does not require a cooling mechanism, so that the system configuration of optical communication systems can be made simple and inexpensive.

Further, there are high expectations for semiconductor QDs in the application of future information communication technologies such as quantum information processing and quantum communication. As discussed before, QDs have a three-dimensional confinement structure, and many research groups all over the world are now attempting to perform quantum information processing by using confined excitons—combined states of electrons and holes—in this confinement structure and by applying coherent control to the excitons. Attempts are also underway at applying QDs to interception-free quantum communication by controlling each of the photons generated by QDs [2]. The advantage of implementing these processes with semiconductor QDs is that such advances will enable the production of smaller devices as compared to other methods.

In addition, the application of QDs to nanophotonic devices (the subject of this book) is expected [3]. The fabrication of a new type of photonic device with low power consumption can be expected by applying the peculiar interaction of the optical near-field, because the QD size is less than the wavelength of light. In this case, the size and fabrication site of the QD should be controlled accurately, which is a challenging research area. In this chapter, we will discuss the fabrication method of semiconductor QDs, which is an important fundamental technology to realize nanophotonic devices. In the first section, we will introduce a self-assembled QD and discuss the control of density and emission wavelength of QDs. In the second section, we will discuss how to control the position of QDs. In this section, we will show the selective-area growth technique to control the position of QDs.

4.2 Fabrication of Self-Assembled QDs

Quantum structures including QDs can be fabricated through the manipulation of energy bandgaps in semiconductors. Namely, by fabricating a structure in which a material with a larger energy gap surrounds a material with a smaller energy gap, electrons and holes are confined in the material with the smaller energy gap. However, to obtain sufficient quantum effects, these structures must be on the order of several tens of nanometers or even smaller. We require extremely precise techniques for fabricating arbitrary nanostructures. Recently, advanced semiconductor crystal growth technology has led to a mature technology for fabricating semiconductor films at a precision of 1 nm or less based on molecular beam epitaxy (MBE) and other techniques. Therefore, we can easily fabricate QWs, the one-dimensional carrier confinement structure. These QWs can be fabricated by precisely controlling the growth rate and growth time in semiconductor crystal growth. This technique has led to the realization of semiconductor lasers that operate at room temperature, and these lasers have been applied to various devices, including those aimed at many applications.

In contrast to the thin-film deposition technology for fabricating QWs, QDs are more difficult to fabricate. As QDs require a confinement structure in all the three dimensions, we now require a structure fabrication technology that can allow processes on the order of several tens of nanometers or less along directions parallel to the surface. The first attempt at fabricating QDs involved the formation of a QW, patterning of the well with electron-beam lithography equipment, and etching of the pattern. However, this top-down method presents problems in that it cannot yield high-quality QDs, since it damages the sample during etching; additionally, it cannot yield high-density QDs.

In the early 1990s, a new method of QD fabrication was invented that made use of the self-assembling nature of semiconductor crystal growth. In crystal growth in a lattice-mismatched material system, structures are self-assembled at sizes on the order of several tens of nanometers along directions parallel to the surface. The QDs obtained in this manner are referred to as self-assembled QDs [4]. Normally, in the crystal growth of a lattice-mismatched material system, defects and dislocations are formed to relax the strain energy of the growth film when this film cannot withstand the lattice strain. The crystal quality deteriorates because of defects and dislocations, and thus materials with low lattice mismatching are generally selected for semiconductor crystal growth. On the contrary, self-assembled QDs make favorable use of this strain in a lattice-mismatched system. One of the most popular material systems for self-assembled QDs consists of a combination of GaAs and InAs. In this combination, InAs forms the QDs. GaAs and InAs have lattice constants of 5.653 Å and 6.058 Å, respectively, and the lattice mismatch between them is approximately 7%. When InAs is grown on GaAs, InAs first grows two-dimensionally and moves to three-dimensional growth when the layer exceeds approximately 1.5 monolayers (one monolayer corresponds to half the lattice constant.). This kind of growth mode is referred to as the Stranski–Krastanow mode (S–K mode) [5]. This three-dimensional growth forms InAs island structures (InAs QDs) on the sample surface, each with a diameter of approximately several tens of nanometers. When InAs is grown without being stopped, defects and dislocations are formed in the crystal, as discussed earlier, and the crystal quality deteriorates. However, defects and dislocations do not arise just after the formation of the InAs QDs; if the growth of InAs is stopped at the appropriate growth amount and the sample is embedded with GaAs (or another material with a larger bandgap than that of InAs), InAs QDs can be successfully produced. As an example, Figure 4.2 shows an atomic force microscope (AFM) image of the self-assembled InAs QDs fabricated on GaAs with MBE in our group facilities. The fabrication procedure is as follows. Firstly, the GaAs substrate was thermally cleaned by holding it in a growth chamber at 610 °C. After cleaning, a 150-nm thick GaAs buffer layer was grown at a rate of 1 ML/s at 580 °C. A 2-ML InAs layer was then grown at 0.04 ML/s to fabricate the self-assembled QDs. In this sample, QDs with an average diameter of 40 nm, average height of 9 nm, and density of $2.5 \times 10^{10}/cm^2$ were obtained. It is the characteristic of the self-assembling method that high-quality QDs are obtained without damage, as this is a complete vacuum-based process; additionally, the obtained QDs feature a higher density than that obtained from the top-down fabrication method. By using this type of QDs, diverse applications are now under development. Among these, the use in optical communication devices is just

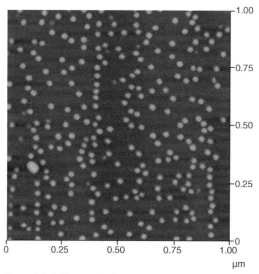

Figure 4.2 Self-assembled InAs QDs grown on GaAs substrate.

a step away from practical application. In particular, the QD laser in the 1.3-μm band and the QD semiconductor optical amplifier (SOA) operate at a lower threshold current, depend less on temperature, and offer better high-speed signal processing performance than lasers and optical amplifiers based on QWs [6–13].

4.2.1
Control of Density and Emission Wavelength of Self-Assembled QDs

It is very important to control the density and emission wavelength of QDs for their application to nanophotonic devices using optical near-fields. In controlling the density of QDs, a fabrication technique for low-density QDs is required because we need to distinguish between individual QDs in the scanning near-field optical microscope (SNOM) observations [14]. The size of the probe aperture of SNOM is usually around 50–100 nm. Therefore, it is necessary that the distance between neighboring QDs be more than 50 nm to distinctly distinguish between individual QD emissions.

In addition, a photodetector with high sensitivity and low noise is required to obtain signals at the output of nanophotonic devices because the signal from a single QD is usually very weak. A photomultiplier tube (PMT) and a Si-based avalanche photodiode (Si-APD) satisfy these requirements; these devices enable the detection of emissions from single QDs. These high-performance devices can usually detect emissions with wavelengths shorter than 1 μm. Therefore, we need to develop a QD fabrication technique with emissions below 1 μm so that these high-performance photodetectors can be used. On the other hand, when it is required to introduce the signal from a QD in nanophotonic devices into optical fibers, the emission wavelength of the QD should be 1.3 or 1.55 μm, which is suitable for fiber-optic communication systems. In this case, an amplification technique for weak signals from single QDs is needed, as well as controlling the emission wavelength. Therefore, we need a fabrication technology that can arbitrarily control the density and emission wavelength of QDs in order to fabricate QDs for nanophotonic devices (Figure 4.3). In this subsection, we will discuss how we control the density and wavelength of self-assembled QDs by changing the materials and fabrication processes.

4.2.2
Shortening Emission Wavelengths of Self-Assembled QDs

Figure 4.4 shows the photoluminescence (PL) spectrum of the InAs QDs measured at 12 K. The sample was fabricated by embedding the QDs with a GaAs layer, as shown in Figure 4.2. The PL measurement system consists of a 532-nm second-harmonic line of an Nd:YVO$_4$ laser excited using a semiconductor laser diode, 250-mm monochromator, and electrically cooled PbS photodetector. The PL spectrum shows an emission peak at 1050 nm. At room temperature, this spectrum peak shifts to around 1150 nm. Therefore, the fabrication technique that permits QD emissions to shorter wavelengths will be required when using high-performance photodetectors

74 | *4 Fabrication of Quantum Dots for Nanophotonic Devices*

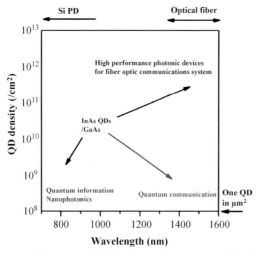

Figure 4.3 Demands for density and emission wavelength of self-assembled QDs.

such as PMT or Si-APD (mentioned earlier). In GaAs and InAs material systems, the incorporation of Al makes their bandgaps larger, thereby shortening the emission wavelength. Therefore, the emission of InAs QDs at 1050 nm grown on GaAs can be shifted to a shorter-wavelength region if we incorporate Al in both the QDs and the GaAs barrier. Consequently, we first fabricated InAlAs self-assembled QDs embedded by AlGaAs. The schematic sample structure is shown in Figure 4.5(a). The GaAs (001) substrate cleaning and the growth of a GaAs buffer layer are carried out under the same condition as that in the case of InAs/GaAs QDs. After the growth of the

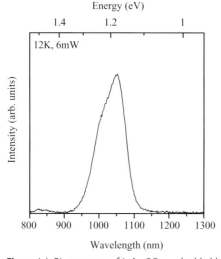

Figure 4.4 PL spectrum of InAs QDs embedded by a GaAs cap layer measured at 12 K.

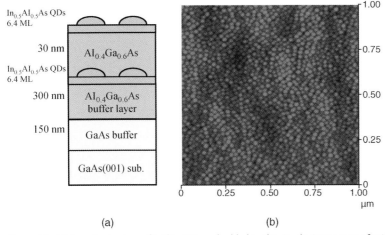

Figure 4.5 (a) Sample structure of InAlAs QDs embedded in AlGaAs. (b) AFM image of InAlAs QDs.

GaAs buffer layer, a 300-nm thick $Al_{0.4}Ga_{0.6}As$ layer was grown at 580 °C at a growth rate of 1.0 ML/s. After the growth of the AlGaAs layer, the substrate temperature was decreased to 500 °C, and a 6.4-ML thick $In_{0.5}Al_{0.5}As$ QD layer was grown at a growth rate of 0.08 ML/s. After the growth of InAlAs QDs, a 30-nm thick $Al_{0.4}Ga_{0.6}As$ cap layer was grown at a growth rate of 0.1 ML/s for the PL sample. For the AFM measurement, additional QD layers were grown. Figure 4.5(b) shows the AFM image of InAlAs QDs. In this sample, high-density InAlAs QDs were fabricated. The average diameter and height of InAlAs QDs are 28.3 nm, and 2.3 nm, respectively. The density of QDs is $1.5 \times 10^{11}/cm^2$, which is about 5 times larger than that of InAs QDs grown on GaAs. The increase in QD density originates from a decrease in the diffusion length of group-III atoms by incorporating Al atoms in the QD layer. This is consistent with the fact that Al has a lower vapor pressure than In. In addition, the existence of Al atoms in the layer lying under the QD layer decreases the diffusion length of the QD materials. The latter effect will be discussed in the next section. The PL spectrum of the InAlAs QDs measured at 12 K is shown in Figure 4.6(a). The emission of InAlAs QDs was clearly measured at 658 nm. In addition, the emission from a single InAlAs QD was observed in the SNOM measurement at a lower-energy position of the PL spectrum, as shown in Figure 4.6(b) [15]. In this case, we could observe the single QD emission because only a few QDs have these energy levels, whereas the total density of QDs is very high in this sample.

As mentioned before, the InAlAs QDs emit in the shorter-wavelength region as compared to the InAs QDs embedded in GaAs. However, the density of InAlAs is too high to distinguish individual QD emissions even if we use the SNOM measurement system. Fortunately, we can observe the single-QD emission at lower-energy positions of the macroscopic PL spectrum where the number of QDs with this energy level is lower than that with an energy level around the PL peak. The fabrication technique to achieve lower QD density is required when we use individual QDs for nanophotonic devices, which have energy levels over the entire PL spectrum range.

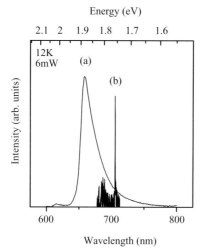

Figure 4.6 PL spectrum of InAlAs measured at 12 K. (a) Far-field measurement (QDs ensemble) and (b) near-field measurement (single QD).

The diffusion length of group-III atoms should be large to form low-density QDs. From this viewpoint, we used InAs for fabricating the QD materials, and embedded them in AlGaAs. When the QD material is changed from InAlAs to InAs, the bandgap of the QDs becomes small, i.e. the emission wavelength of the QDs becomes larger. We increase the Al composition of the AlGaAs barrier layer to suppress the elongation of the emission wavelength of the QDs. Figures 4.7(a) and (b) show the schematic sample structure and AFM image of the sample, respectively. As shown in the AFM image, the distance between the neighboring QDs becomes large, which

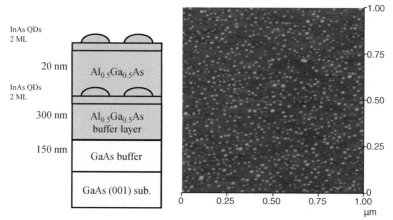

Figure 4.7 (a) Sample structure of InAs QDs embedded in AlGaAs. (b) AFM image of InAs QDs grown on AlGaAs.

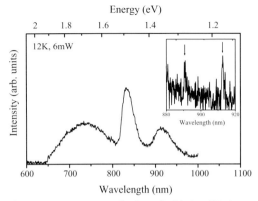

Figure 4.8 PL spectrum of InAs embedded in AlGaAs measured at 12 K. Inset shows near-field spectrum measured by SNOM. The arrows show the emissions from single QDs.

shows the tendency of decreasing QD density. The average QD diameter, height, and density are 24.5 nm, 2.9 nm, and $9.5 \times 10^{10}/cm^2$, respectively. As compared to InAlAs QDs, the size of the QD becomes smaller and the density decreases when using InAs for the QDs. However, the number of QDs in the SNOM aperture is about 8. Therefore, a growth technique that yields lower-density QDs is required. The high density of QDs originates from the small diffusion length of group-III atoms by incorporating Al atoms in layers lying under the QDs. The existence of Al atoms in the underlying layer decreases the diffusion length of the QD materials because the chemical bonding of Al-related material is strong. Therefore, a diffusing In atom can be captured in a stable site easier on an AlGaAs surface than that on a GaAs surface. The PL spectrum of this sample is shown in Figure 4.8. The emission at 820 nm originates from the GaAs bulk. The emissions around the GaAs peak are considered to be the QD emissions. In this sample, it is difficult to clearly separate the GaAs and QD emissions using the macroscopic PL measurements. In the SNOM measurements, the emission from a single QD was observed in a longer-wavelength region, as shown in the inset of Figure 4.8. However, the emission efficiency is not so high and therefore the signal from a single QD is very weak. We need to improve the growth technique to achieve QDs with low density and high emission efficiency.

4.2.3
Controlling the Density of Self-Assembled QDs

Up to now, we have discussed a fabrication technique of self-assembled QDs permitting emissions below 1 μm. However, it is difficult to fabricate low-density QDs as low as a single QD in a 100-nm² region with material systems that contain Al. The advantage of using materials that contain Al is that a shorter-wavelength emission can be obtained and its emission can be detected by a Si photodiode. However, high-performance photodetectors made of InGaAs and so on have been recently developed, which can detect longer-wavelength emissions as compared to those detectable by Si photodetectors. Although Si photodetectors have considerably better processes and noise characteristics, these high-performance photodetectors have improved sensitivity since emissions from single QDs can be detected. In this section, we will discuss a technique with which QD density can be decreased, especially in the fabrication of InAs QDs grown on GaAs and antimonide QDs to yield low-density QDs.

As discussed before, the density of InAs QDs grown on GaAs under ordinary conditions is $2.5 \times 10^{10}/cm^2$. If we can fabricate a device with dimensions of 100 nm × 100 nm, several QDs can be fabricated on a single device. The use of material systems employing Al induced the formation of high-density QDs because the diffusion length of group-III atoms of QDs became small. Therefore, the enhancement of the surface diffusion of group-III atoms should decrease the density of QDs. The enhancement of the surface diffusion length can be achieved to increase the growth temperature of QDs or to decrease the growth rate of QDs. Firstly, we carried out the high-temperature growth process. The growth condition of the sample is the same as that in Figure 4.2, except for the growth temperature of QDs. Two layers of InAs QDs separated by the GaAs layer were grown at 520 °C. Figure 4.9 shows the AFM image of this sample. Low-density uniform InAs QDs with a slightly larger size as compared to that grown at 500 °C were formed in this sample. The density of QDs reduces to $5.0 \times 10^{10}/cm^2$. The average diameter and height are 55.3 and 14.4 nm, respectively. These results indicate that the control of the growth temperature of InAs QDs is important to control the QD density. Figure 4.10 shows the PL spectrum of this sample measured at 12 K. The emission peaked at 1061 nm with a full width at half-maximum (FWHM) value of 48.7 meV corresponding to the QD ground-state transition. The emissions at 920 and 830 nm correspond to the emissions from the InAs wetting layer and GaAs bulk, respectively. The emission wavelength of InAs QDs in this sample was larger than that of the sample embedded by AlGaAs. Although the density of InAs QDs became lower, the emission from the QDs became

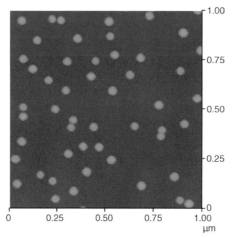

Figure 4.9 AFM image of InAs QDs grown at 520 °C.

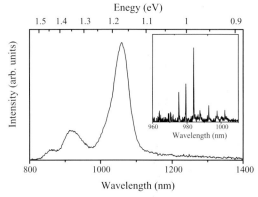

Figure 4.10 PL spectrum of InAs QDs grown at 520 °C. Inset shows the near-field spectrum measured by SNOM, which shows several emissions from single QDs.

stronger than those from the sample with AlGaAs. This can be attributed to the reduction in the carrier recombinations at the nonradiative recombination center because of Al. The inset of Figure 4.10 shows the near-field spectrum measured by SNOM. Several emissions from single QDs are also observed in this sample. Although the density of QDs decreased even when Al was not used, more emissions were observed from single QDs. This also indicates that the emission efficiency was improved in this sample.

Next, a growth process affording a low growth rate was carried out. Since the low growth rate also enhances the diffusion of group-III atoms, the formation of low-density QDs was expected. However, since the growth rate for the sample in Figure 4.2 already had a minimum of 0.04 ML/s, it is difficult to reproducibly fix a lower growth rate than this value. To overcome this problem, we used a method of repeating the growth and growth interruption to obtain a practical low growth rate. The shutter sequence is shown in Figure 4.11 to grow InAs QDs. An In flux value of 0.04 ML/s was used in this sequence. Here, In was supplied to the substrate for 2 s, followed by growth interruption for 8 s. This growth sequence corresponds to a growth rate of 0.008 ML/s. A 2-ML thick InAs QD layer was grown by repeating this sequence 25 times. The other growth conditions were the same as the sample in Figure 4.2. Figure 4.12 shows the AFM image of InAs grown with a shutter sequence.

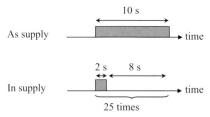

Figure 4.11 Shutter sequence to achieve ultralow growth rate.

4 Fabrication of Quantum Dots for Nanophotonic Devices

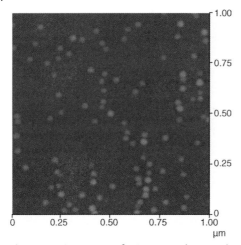

Figure 4.12 AFM image of InAs QDs with a growth rate of 0.008 ML/s.

A lower density of QDs than that of the sample shown in Figure 4.2 is observed. The averaged diameter, height, and density of this sample are 40.5 nm, 5.1 nm, and $1.0 \times 10^{10}/\mathrm{cm}^2$, respectively. The lower density of InAs QDs is caused by the enhancement of the surface diffusion of group-III atoms during the growth of QDs. Although there are many QDs with these average dimensions, some smaller QDs, which had probably just formed, are also found. It indicates that the distribution of QD dimensions has bipolarity, which is prohibitive for the control of QD dimensions. Therefore, the control of the growth temperature is more suitable for the control of QD density because of its simplicity of growth mechanisms.

4.2.4
Fabrication of Self-Assembled QDs with Antimonide-Related Materials

In this subsection, the fabrication of antimonide QDs will be discussed. The incorporation of antimony (Sb) induces surfactant effects during the growth of GaAs/AlGaAs QWs. It improves the optical properties such as PL intensity [16–18]. In addition, there are some effects with regard to the suppression of generating defects and dislocations in the crystal growth with lattice-mismatched semiconductor material systems [19–21]. There are various properties in the surfactant effects: the enhancement of two-dimensional growth is one example. Therefore, the growth of antimonide QDs on GaAs substrates may reveal a different situation even if they have a similar amount of lattice mismatch. Actually, the formation of lower-density antimonide QDs will be expected because the two-dimensional growth should be enhanced by the Sb surfactant effect. Therefore, we tried to fabricate InGaSb QDs on a GaAs substrate. The conditions of substrate cleaning and growth of the GaAs buffer layer were the same as those during the growth of InAs QDs. After the growth of the GaAs buffer layer, a 2-ML thick $In_{0.5}Ga_{0.5}Sb$ QD layer was grown at 400 °C. The Sb flux during the growth of InGaSb QDs was 3.8×10^{-7} Torr. Figure 4.13 shows

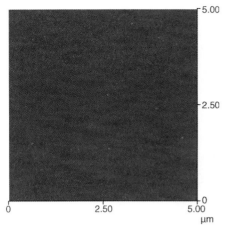

Figure 4.13 AFM image of InGaSb QDs on GaAs.

the AFM image of the InGaSb QDs grown on the GaAs substrate. The image was measured in a $5 \times 5\,\mu m$ region. Only 10 InGaSb QDs were observed in this region, which corresponds to a density of $5.2 \times 10^7/cm^2$. The average diameter and height of the InGaSb QDs are 80 and 8 nm, respectively. This result indicates that extremely low density QDs can be fabricated by using InGaSb as the QD material.

During the fabrication of self assembled InGaSb QDs, extremely low density QDs as compared to InAs QDs can be obtained. These low-density QDs have an advantage during the SNOM measurements of single QD emissions. However, there is less flexibility for the fabrication of nanophotonic devices that employ several QDs and integration of nanophotonic devices because the density of QDs is too low. Ideally, the technology to control the density of QDs arbitrarily from the order of $10^7/cm^2$ to $10^{10}/cm^2$ should be established. Although it is considered that the density of InGaSb QDs can be controlled by changing the substrate temperature or growth rate during the InGaSb QD growth, we used a new technique of Si irradiation as the antisurfactant. The antisurfactant effect exhibits an opposite effect to the surfactant effect, such as the enhancement of three-dimensional growth. The antisurfactant effect of Si has been shown in the growth of GaN on AlN [22]. Therefore, it is expected that Si irradiation before the growth of InGaSb QDs on GaAs also acts as the antisurfactant. The Si irradiation was carried out before the growth of InGaSb QDs under the same conditions as that of the sample mentioned in Figure 4.13. The irradiation rate of Si atoms was $1.4 \times 10^{10}/cm^2/s$, which was estimated by the carrier density of Si-doped GaAs thin films. Figure 4.14 shows the AFM image of InGaSb QDs with Si atom irradiation for 30 s, which corresponds to the deposition of $4.2 \times 10^{11}/cm^2$ Si atoms. We found that the density of InGaSb QDs is drastically increased with Si atom irradiation. High-density InGaSb QDs ($\sim 4.4 \times 10^9/cm^2$) can be fabricated, corresponding to a density of about 100 times higher than that without the irradiation. Figure 4.15 shows the dependence of dot density of InGaSb QDs on the Si atom irradiation times. The density of InGaSb QDs increases by increasing the duration of Si atom irradiation and the QD density begins to saturate over 30 s.

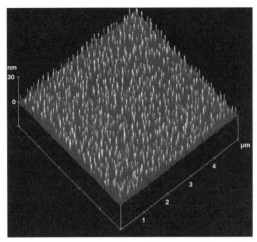

Figure 4.14 AFM image of InGaSb QDs on GaAs with Si atom irradiation.

We confirmed these saturation characteristics up to 90 s. Therefore, we achieved the growth of a high-density InGaSb QD layer by irradiating Si atoms of approximately $4.2 \times 10^{11}/cm^2$ onto a GaAs surface. PL measurements were carried out for the sample with a cap layer. Figure 4.16 shows the PL spectra of InGaSb QDs grown with and without irradiation measured at room temperature. Although we found that the PL intensity is drastically enhanced by Si atom irradiation, we also observed a weak PL intensity in the QD samples without irradiation. This weak PL from QDs without irradiation is caused by the low-density InGaSb QDs (Figure 4.13). Therefore, the enhanced PL intensity depends on the increasing QD density. The PL from QDs with Si atom irradiation has a broadband spectrum of 0.25 eV (FWHM) due to the wide distribution of the QD dimensions. A PL peak wavelength of around 1.18 μm was observed for both the samples. In addition, large intensities for the longer

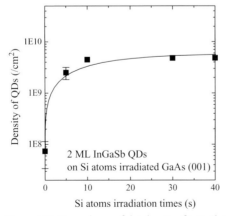

Figure 4.15 Dependence of dot density of InGaSb QDs on Si atom irradiation times.

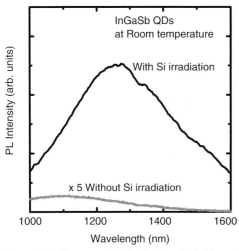

Figure 4.16 PL spectra of InGaSb QDs with/without Si atom irradiation.

wavelengths around 1.3 μm can also be observed. These results indicate that InGaSb QDs on GaAs are candidate materials for the fabrication of novel photonic devices for fiber-optic communication networks [23,24]. On the other hand, since the emission wavelength of InGaSb QDs is larger than that of InAs QDs, there are a few high-performance photodetectors that can yield low-noise high-speed responses for nanophotonic devices. However, after the signals from nanophotonic devices are amplified, the signals at 1.3 μm can be directly connected to fiber-optic networks.

4.2.5
Fabrication of Ultrahigh-Density QDs

We have already discussed the control of density of self-assembled QDs, particularly the development of a fabrication technique for low-density QDs, from the viewpoint of detection and control of emissions from single QDs using SNOM. In this subsection, we will discuss the fabrication of high-density self-assembled QDs. The principle of nanophotonic devices lies in the control of excitations in a single or several QDs by using optical near-field interactions. To obtain the output from nanophotonic devices after performing some processing, the output signal from the nanophotonic devices needs to be amplified because the signals obtained from nanophotonic devices are very small. Therefore, an amplifier for small signals (e.g., SOA) is needed at the exit port of nanophotonic devices. In the SOA, the QDs function as the gain medium; therefore, the fabrication technology for denser QDs represents a key technology for improving device performance. However, even when simply considering increasing the in-plane density, spatial limitations are encountered. For example, assuming that we can fabricate QDs with a diameter of 20 nm in a closely packed structure (Figure 4.17), the surface density is limited to approximately $3 \times 10^{11}/\text{cm}^2$. Further, it is difficult to fabricate such a structure. When one

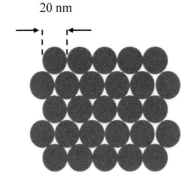

Unit cell: $20 \times 10\sqrt{3} \times \frac{1}{2}$

Quantum dot: 0.5

$A(1QD) = 20 \times 10\sqrt{3} \times \frac{1}{2} \div 0.5 = 346\, nm^2$

Density $= 1/A = 2.89 \times 10^{11} / cm^2$

Figure 4.17 Estimation of density limit when QDs are formed in the closely packed structure.

pursues a denser QD structure, layers comprising QDs can be stacked. However, QD fabrication based on the S–K mode uses the strain energy of the lattice-mismatched material system as the driving force in QD formation. Hence, when the density of QDs is to be increased by stacking, strain accumulation creates a problem. This accumulation may lead to problems such as changes in the size and shape of QDs; further, an excessive accumulation of strain generates defects and dislocations; therefore, the number of stacked layers is generally limited to 10 or less.

To resolve this problem, we developed a strain-compensation method when stacking QDs; this approach enables the stacking of multilayer QDs. Figure 4.18 shows a schematic diagram of this method. We used an InP(311)B substrate for

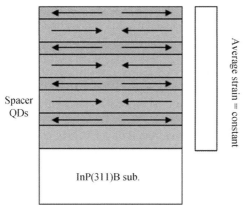

Figure 4.18 Schematic diagram of strain compensation.

Table 4.1 Lattice constants of AlAs, GaAs, InP, and InAs.

Material	AlAs	GaAs	InP	InAs
Lattice constant	5.661 Å	5.6533 Å	5.8687 Å	6.0583 Å

fabricating the QDs. As shown in Table 4.1, the lattice constant of InP is between that of GaAs or AlAs and InAs; as a result, various material systems can be grown on an InP substrate. In the present study, we have devised a structure in which the InAs QDs fabricated on InP are embedded in InGaAlAs, which has a slightly smaller lattice constant than InP. In this manner, the tensile strain generated in InAs is compensated by the compressive strain in InGaAlAs, which solves the problem of accumulated strain energy when fabricating the stacked structure. We determined the conditions for strain compensation based on the following equations:

$$d_{QD} \cdot \varepsilon_{QD} = -d_s \cdot \varepsilon_s$$
$$\varepsilon_{QD} = (a_{InAs} - a_{InP})/a_{InP}$$
$$\varepsilon_s = (a_s - a_{InP})/a_{InP}$$

where d_{QD} and d_s are the film thicknesses of the QD layer and strain-compensation layer, respectively. Further, a_{InAs}, a_{InP}, and a_s are the lattice constants of InAs, InP, and InGaAlAs strain-compensation layers, respectively; ε_{QD} and ε_s are the amounts of strain with respect to the InP substrate in the InAs QDs and strain-compensation layer, respectively. The sample is prepared as follows. The InP(311)B substrate is placed in an MBE growth chamber and thermally cleaned at 500 °C for 10 min to produce a clean surface. Then, a 150-nm thick lattice-matched InAlAs buffer layer is grown. Finally, the InAs QDs and InGaAlAs strain-compensation layer are grown alternately, yielding the stacked structure.

First, to verify the effect of strain compensation, the stacked InAs QD structure was fabricated on the InP(311)B substrate. Figure 4.19 shows the AFM measurement results. In this figure, (a) shows the surface morphology of the single-layer InAs QDs on the InP(311)B substrate, and (b) shows a sample stacked with 20 layers of InAs QDs with 10-nm InGaAlAs spacer layers, which satisfies the strain-compensation conditions. Uniform QDs are also formed for the single-layer InAs QDs on the InP substrate, and strain compensation maintains uniformity in the size distribution of QDs after stacking 20 layers. These results clearly indicate that strain compensation plays an important role in stacking multilayer QDs. As the QD distribution is more uniform than that without strain compensation, we may further conclude that this method also suppresses the formation of defects and dislocations. It is also known that adding Al to the strain-compensation layers suppresses In surface segregation. It is, therefore, important to use InGaAlAs containing Al in the intermediate layers in order to form a uniform stacked structure of QD layers [25].

To confirm the absence of dislocations, we also observed a sample cross-section with a scanning transmission electron microscope (STEM). Figure 4.20 shows the results. In this sample, 30 layers are stacked with 20 nm of InGaAlAs

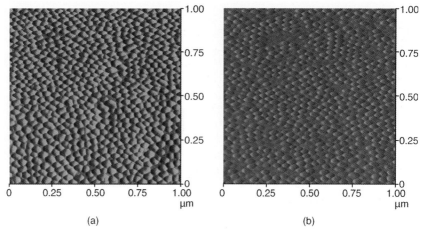

Figure 4.19 Comparison of QD shapes associated with stacking: (a) single-layer InAs QDs on InP, and (b) 20-layer stacked InAs QDs on InP.

strain-compensation layers. The inset shows the result of the AFM measurements for the sample surface. Uniform QDs have formed on the sample surface, and cross-sectional STEM measurements do not indicate the formation of dislocations. Thus, it is clear that the creation of a stacked structure with strain compensation not only maintains the uniformity of QDs but also suppresses the formation of dislocations, leading to the creation of high-quality QDs. The advantage of stacked structures based on the strain-compensation method lies in the fact that the number of stacked layers is, in principle, unlimited as long as the strain-compensation condition is satisfied. Hence, stacking can be repeated many times, enabling the fabrication of extremely dense QDs. The present study led to the fabrication of a stacked structure

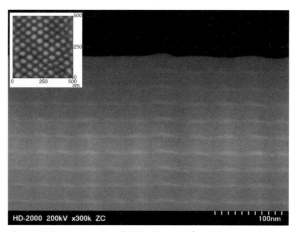

Figure 4.20 Cross-sectional STEM image of a 30-layer stacked structure.

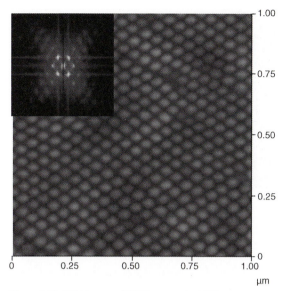

Figure 4.21 AFM image of 150-layer stacked QDs.

of up to 150 layers. Figure 4.21 shows the surface morphology of the InAs QDs with a stacking of 150 layers. Despite the extremely large number of stacked QD layers, the surface morphology shows no degradation. Figure 4.8 shows the dependence of size and density of QDs on the number of stacked layers. The diameter, height, and density of the QDs change slightly as a function of the number of stacked layers, which indicates the effectiveness of the strain-compensation method in fabricating a stacked structure. The cross-sectional STEM measurement discussed above also supports the maintenance of QD size distribution. Figure 4.22 shows that approximately $4 \times 10^{10}/cm^2$ QDs are formed per layer. Thus, over $5 \times 10^{12}/cm^2$ QDs are formed in a sample with 150 stacked layers, a result that would prove to be impossible

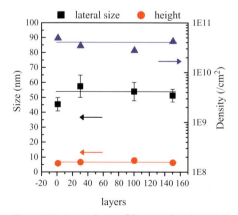

Figure 4.22 Dependence of diameter, height, and density of QDs on the number of stacking layers.

when using conventional methods. Thus far, this number of stacked QD layers has never been achieved. The inset (Figure 4.21) shows a two-dimensional fast Fourier-transform (2DFFT) image. This 2DFFT image clearly reveals the higher-order satellite peaks. As is evident from the AFM image, this indicates a two-dimensional array structure of QDs. The symmetry of the 2DFFT shows that the QDs are formed in a closely packed structure with sixfold symmetry. The formation of the QD array as the number of stacked layers increases is probably due to the redistribution of strain after the formation of each strain-compensation layer. In other words, although the strain energy is counterbalanced and prevented from accumulating in the material system as a whole, a nonuniform strain is distributed on the sample surface after the growth of a strain-compensation layer due to the variances in the positions of the embedded QDs. The lattice constant is slightly larger directly above the QDs, yielding the dominant sites for the generation of QDs in the next layer. This phenomenon can also be confirmed in the cross-sectional STEM measurement, which shows that the QDs in the next layer are formed above the QDs of the lower layer. As the strain-compensation condition is satisfied by necessity, the lattice constant is slightly smaller at positions other than those directly above the QDs, compensating for the tensile and compressive strains in the entire system. With regard to a single QD, it is evident that the QD in the next layer is formed above the QD of the lower layer, continuing the morphology of the first layer. However, the fact that an array formation is promoted as the number of stacking layers increases indicates that the strain distribution formed on the spacer layer is due to the strain interactions involving the QDs in the nearby area, as well as the QDs directly below. Thus, the formation of a strain field involving multiple dots is important in the formation of QD arrays. It is easily deduced that the propagation of the strain field strongly depends on the thickness of the intermediate layer. In other words, thinner intermediate layers will lead to a stronger effect of QDs directly below in relation to the surrounding QDs, so the system will continue the morphology of the first layer as is. Thicker intermediate layers involve an extremely large number of QDs in the generation of the strain field; therefore, the strain field is averaged and the QD array formation may disappear. It should also be noted that the (311)B substrate surface tends to cause the formation of QD arrays [26]. This is also considered to lend additional strength to the formation of a QD array structure. These phenomena have been well studied with respect to the PbSe/PbEuTe material system [27]. To investigate this aspect, the present study also involved the fabrication of samples stacked with 150 layers of 10- and 60-nm strain-compensation layers. Figure 4.23 shows the AFM observation results. Here, (a) and (b) show the results with the 10- and 60-nm strain-compensation layers, respectively. As expected, the array structure eventually dissipates with both thick and thin strain-compensation layers. Thus, it is clear that the film-thickness control of the strain-compensation layer is an important element of QD array formation. As discussed at the outset, QDs have discrete energy levels with a delta-function-form density of states. As the energy can be manipulated, the QDs are sometimes likened to artificial atoms. If we can fabricate a three-dimensional array structure of QDs based on stacked QDs, we can obtain an array structure of artificial atoms—in other words, artificial crystals. Here, we consider that the following three factors will be of

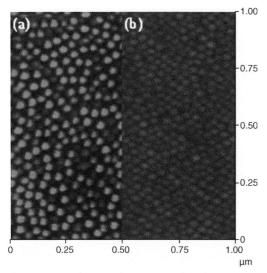

Figure 4.23 150-layer stacking structure based on (a) 10-nm spacer layer and (b) 60-nm spacer layer. (b) 150-layer stacking structure based on (a) 10-nm spacer layer and (b) 60-nm spacer layer.

particular importance: stacking based on strain compensation, QD array formation based on thickness control of the intermediate layers, and control of coupled states between the QD layers.

Finally, we will discuss the optical properties of the stacked QDs. Strain-compensation layers of 20 nm are used and the PL of the sample stacked at 150 layers is measured at room temperature. To excite the sample, the 532-nm second harmonic of a diode-pumped Nd:YVO$_4$ laser is used. A 250-mm monochromator and an electrically cooled PbS photodetector are used for spectroscopy and detection of the emitted light. Figure 4.24 shows the measurement results. The figure shows strong emission even at room temperature. These results also demonstrate the effectiveness of suppressing dislocation by strain compensation and enhancing the emission intensity using increased density. This PL measurement shows a spectrum with its main peak near approximately 1.5 μm and a shoulder structure on the higher-energy side. Based on a simple calculation of quantum levels, these peaks are found to agree with the energy levels of the ground state, first excited state, and second excited state of the QDs. The corresponding values are indicated in the figure. The FWHM value of the ground state is approximately 40 meV. The emission wavelength of this sample corresponds to that of fiber-optic communication and can be expected to be applied to QD lasers and SOAs. The density of QDs is particularly important in QD lasers and SOAs, because it is the source of gain. Thus, the fabrication technology for high-density QDs used in the present study has the potential to improve the performance of conventional semiconductor devices by a significant amount. Further, InAs QDs on GaAs are subject to large compressive stress caused by GaAs, resulting in the formation of a large bandgap and a problematic restriction of

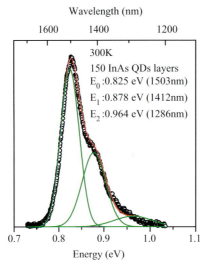

Figure 4.24 PL spectrum of a 150-layer stacking structure.

emission to the 1-μm band. However, as the InP substrate has a larger lattice constant than GaAs, the compressive stress applied to InAs on an InP substrate is smaller; accordingly, we have obtained emissions of the InAs QDs in the fiber-optic communication wavelength. These emission characteristics can be modified by changing the volume of InAs QDs and barrier heights of the intermediate layers. If we combine QDs with various emission wavelengths and increase the density while satisfying the strain-compensation condition, we believe that it will be possible to produce SOAs with higher efficiency and broader bandwidth than those available with current products.

4.2.6
Summary

We fabricated self-assembled QDs using various growth conditions and materials to control the density of QDs and the emission wavelength with MBE. The types of semiconductor compounds and the density and wavelengths of QDs for each compound are shown in Figure 4.25. For InAs QDs grown on a GaAs substrate, the emission wavelength was controlled over a wide range (700–1100 nm) by changing the materials under which the QDs were buried. The density of QDs was reduced to the order of $10^9/cm^2$ by using a low growth rate and high growth temperature. In addition, self-assembled QDs fabricated with antimonide-related material showed a lower density. The density of InGaSb QDs was of the order of $10^8/cm^2$. In particular, the use of Si atom irradiation before QD growth increased the density of InGaSb QDs to the order of $10^{10}/cm^2$. On the other hand, ultrahigh-density InAs QDs were grown on an InP (311)B substrate. In this case, we used strain compensation to stack a large number of QD layers. We successfully stacked up to 150 QD layers, and the total density of the stacked QDs was of the order of $10^{13}/cm^2$.

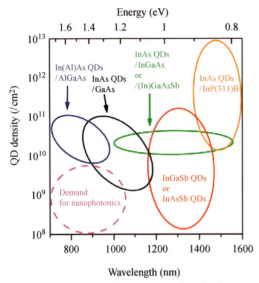
Figure 4.25 Density and emission wavelength of self-assembled QDs.

QDs for lasers used in fiber-optic communication systems have been fabricated using the growth technique mentioned above. However, nanophotonic device applications require advances in technology before they can be successfully developed. For such applications, a fabrication technique is needed for producing QDs that have a sufficiently low density and controllable emission wavelengths. Low-density InAs QDs are used in scanning near-field microscopy to analyze the emissions of individual QDs, which could help lead to the development of new nanophotonic devices. Future work includes finding ways to control the positions of QDs in order to effectively use individual QDs for the previously mentioned devices.

4.3 Fabrication Techniques of Site-Controlled Nanostructures

A site-controlled fabrication technique for a quantum dot measuring several nanometers constitutes a crucial technology in relation to future applications of nanophotonic and electronic devices. This chapter introduces some effective techniques for the fabrication of site-controlled nanostructures such as quantum dots.

4.3.1 Nanopositioning Technique for Quantum Structures with Dioxide Mask [28]

Nanoscale control of optical-gain distributions on a substrate is an important technique for developing novel photonic devices such as nanosized light emitters, detectors, and single-photon emitters [29]. Modulation of the positions of active media and light waves has enabled the realization of greater coupling between the optical gain and electromagnetic fields of light waves in devices [30]. Additionally, high-performance optical communication tools such as vertical-cavity lasers (VCLs) can be produced using Sb-based quantum structures [31,32]. Therefore, a nanopositioning technique for Sb-based III-V semiconductors is proposed to fabricate position-controlled Sb-based quantum structures for developing advanced optical-communication devices. Using this technique, nanopositioned Sb-based quantum structures such as Sb-based quantum dots (QDs) and quantum wells (QWs) can be realized on a wafer surface.

Figure 4.26 shows the process sequence for fabricating nanopositioned Sb-based quantum structures. First, a 30-nm SiO_2 layer was deposited on GaAs and GaSb (001) wafers using a chemical vapor deposition (CVD) or a sputtering method. Next, periodic circular pits with diameters of 200–700 nm were formed on the SiO_2 film by electron-beam lithography and hydrofluoride wet etching. Sb-QDs and multiquantum wells (MQWs) were then fabricated on the surface in the circular pit patterns using molecular beam epitaxy (MBE). The semiconductor surfaces in the pits were prepared using a thermal cleaning process. Nanopositioned Sb-QD structures

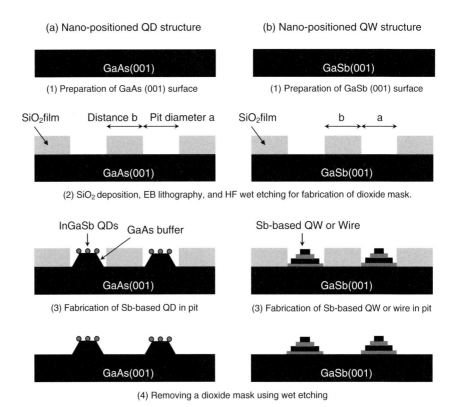

Figure 4.26 Process sequence for fabricating nanopositioned Sb-based quantum structures.

and AlSb/GaSb QW structures can be fabricated on the GaAs and GaSb surfaces, respectively, in the pits.

A 20-nm thick GaAs buffer layer (of growth rate 0.1 ML/s) was formed in the pits at 610 °C using As irradiation (1.2×10^{-5} Torr). The substrate temperature was decreased to 400 °C for QD fabrication. A 2-ML $In_{0.5}Ga_{0.5}Sb$ (0.1 ML/s) was deposited on the surface using Sb-flux irradiation (5.0×10^{-7} Torr). This implies that nanopositioned QD structures can be formed as depicted in the left section of Figure 4.26(a). Figure 4.27 shows a typical atomic-force microscope (AFM) image of an Sb-QD in the pits, wherein the pit diameter and distance are 250 and 150 nm, respectively. The GaAs buffer layer has a pyramidal structure, as shown in Figure 4.27 (b), and the height of the QD structure is approximately 8.8 nm. Additionally, a larger QD was also observed on the two edges of the pyramidal GaAs structure. Several QDs in the pit were observed to have sizes of only a few tenths of a nanometer. Therefore, it is expected that the nanopositioned Sb-QDs will act as single-photon sources for 1.3- and 1.5-μm optical-communication wavebands.

Distribution control of optical-gain materials is expected to emerge as a useful technique for creating novel photonic devices. Additionally, the successful fabrication of a nanopositioned Sb-QW structure was demonstrated in Figure 4.26(b). The GaSb surface in the pits was cleaned at 520 °C using Sb-flux (2×10^{-8} Torr) irradiation. An Sb-QW structure constructed in two-period 25-ML AlSb and 25-ML GaSb layers (0.1 ML/s) was fabricated on the SiO_2-masked GaSb with a 5-nm GaSb buffer layer at 520 °C. Thereafter, the SiO_2 film was removed using a wet etching process. Figure 4.28 shows an AFM image of the nanopositioned AlSb/GaSb QW structure (pit diameter $a = 600$ nm, distance $b = 400$ nm), wherein a QW structure approximately 40 nm high can be observed with nanoscale positioning. Figure 4.29

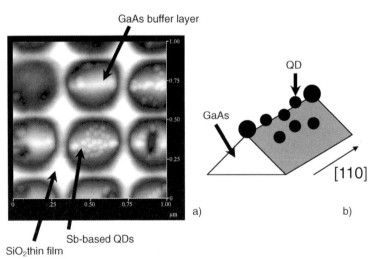

Figure 4.27 (a) AFM image of Sb-based QDs on the pyramidal GaAs in the nanoscaled pits, (b) schematic diagram of QD structures.

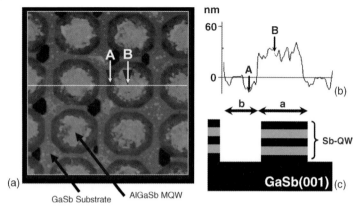

Figure 4.28 (a) Top view and (b) cross-sectional of AFM image of nanopositioned Sb-QW, and (c) schematic image of the nanopositioned QW structure.

shows a photoluminescence spectrum from the nanopositioned Sb-QW structure at RT, exhibiting a peak wavelength at 1512 nm. Additionally, an Sb-based quantum-wire-like structure was formed using this technique when the growth condition of the abovementioned QW structures is altered to a high temperature (560 °C). These results suggest that the layer thickness and aspect ratio of the Sb-QW can be controlled by manipulating the conditions of the nanopositioning technique.

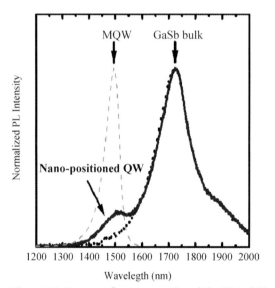

Figure 4.29 PL spectra from nanopositioned Sb-QW (solid line). Also shown are PL spectra from the GaSb surface outside the QW (dotted line) and planar AlSb/GaSb MQWs on GaSb (dashed line).

4.3.2
Artificially Prepared Nanoholes for Arrayed QD Structure Fabrication [33]

A two-dimensional arrayed InGaAs QD structure, with a period ranging from a few tenths to 100 nm, can be realized on a GaAs substrate using a technique for artificially prepared nanoholes. The process sequence of this technique is shown in Figure 4.30 and can be explained as follows. *I.* A 300-nm thick GaAs buffer layer is grown on the GaAs (001) surface in order to form an atomically flat surface. *II.* Circular patterns with a diameter of a few tens of nanometers (approx. 50 nm) are drawn on a resist-coated GaAs surface with a period pattern of a few hundred nanometers (approx. 100 nm) using electron beam (EB) lithography. The drawing region is a 50-micrometer square. *III.* Nanohole etching patterns with a depth of approximately 50 nm are formed on the EB-patterned GaAs surface by chlorine-based reactive ion etching (RIE), and the EB-resist film is removed. *IV.* A 10-nm thin GaAs buffer is regrown on the nanohole-patterned GaAs surface at a low temperature. Subsequently, a 10-monolayer (ML) InGaAs (with an In composition of 0.3) is also deposited on the surface at a low temperature (<400 °C). *V.* The arrayed InGaAs QD structure shown in Figure 4.31 can be realized using an annealing process after InGaAs deposition. From Figure 4.31, a QD density as high as $10^{10}/cm^2$ is realized; this value is comparable to the density of a regular self-assembled QD structure. As shown in Figure 4.32, a multistacked two-dimensional arrayed QD structure is successfully fabricated when a thin GaAs spacer layer with a thickness of several tens of nanometers is further grown on the arrayed QD structure. The thickness of the spacer layer should be very small so that the crystal distortion of a lower arrayed

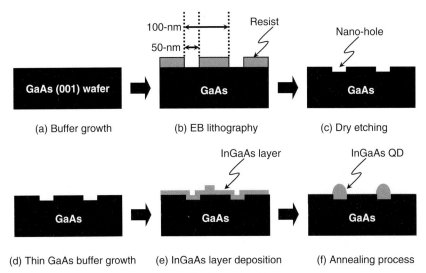

Figure 4.30 Schematic process sequence of technique involving artificially prepared nanoholes for arrayed QD structure fabrication.

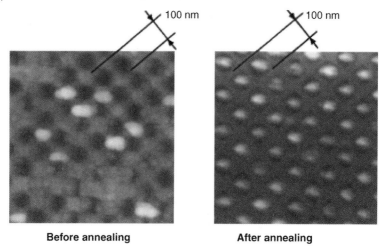

Before annealing **After annealing**

Figure 4.31 Two-dimensional arrayed InGaAs QD structure before and after the annealing process [33].

QD layer may be utilized. The multistacked two-dimensional arrayed QD structure is observed to emit an infrared light (of wavelength approximately 1 micrometer) at room temperature.

4.3.3
Nanojet Probe Method for Site-Controlled InAs QD Structure [34]

It is possible to develop an InAs QD structure at a highly precise arbitrary position on a substrate surface. A nanojet probe (NJP), which is developed by removing the tip probe of an atomic force microscope, is the key device used in this method. Figure 4.33 shows a schematic diagram of a site-controlling technique for a QD structure with an NJP. An In-droplet is formed using the NJP on the atomically flat GaAs (001) surface, after the GaAs buffer layer is grown using the MBE method. The size and volume of the In-droplet can be controlled by applying pulse voltage

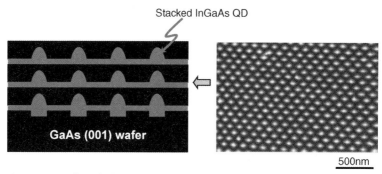

Figure 4.32 Multistacked two-dimensional arrayed QD structure [33].

4.3 Fabrication Techniques of Site-Controlled Nanostructures | 97

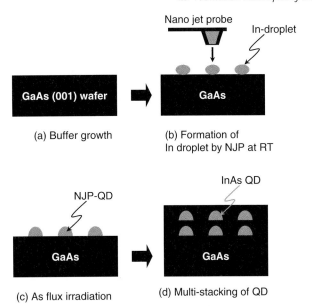

Figure 4.33 Site-controlled QD formation technique using nanojet probe method.

impression to the NJP. The height of a typical In-droplet is approximately 5 nm, and the periodicity length between the In-droplets is approximately 200 nm. To transform the In-droplet into an InAs QD, an annealing process is carried out at approximately 420 °C with As-flux irradiation in the MBE chamber. The multistacked and arrayed InAs QD structure can be fabricated using crystal distortion with a thin GaAs spacer layer; a similar technique was previously described in Section . The spacer layer between the QDs at this point comprises an approximately 10-nm thick GaAs layer. Figure 4.34 shows the surface AFM image of the patterned QD based on the QD template produced by the NJP. The density of the arrayed QDs developed with the NJP is approximately $10^{10}/cm^2$. Moreover, the multistacked InAs QD structure is grown only at a specific location on the NJP-QD structure formed with the NJP, as the

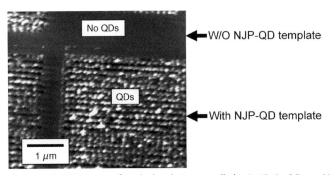

Figure 4.34 AFM image of stacked and site-controlled InAs/GaAs QDs on NJP-QDs template [34].

QD formation is not observed in the region where the NJP-QD is not found. In this method, the formation condition of the accumulating QD on the spacer GaAs surface is influenced by the height of the droplet developed using NJP, including the height of the NJP-QD structure. An upgrade of a precise In-droplet formation using NJP therefore constitutes a significant technological development. In addition, it is confirmed that a room-temperature photoluminescence is observed from the multistacked site-controlled QD structure on the NJP-QD template fabricated using the NJP.

4.3.4
Scanning Tunneling Probe Assisted Nanolithography for Site-Controlled Individual InAs QD Structure [35]

A method involving the use of a scanning tunneling microscope (STM) probe for the fabrication of site-controlled individual InAs QD structures is also proposed. In this method, site controlling of the QD structure is made possible with the use of a nanomask formed on a GaAs surface using the STM probe. Figure 4.35 shows a schematic sequence of the STM-probe-assisted site-controlling technique. The STM probe is brought into contact with the atomically flat surface of the GaAs substrate, and a pulse voltage is applied to the probe in order to create the nanomask on the substrate. Considering that the nanomask is made from a tungsten compound, it can be used steadily under high-temperature conditions, such as during crystal growth in the MBE chamber. After the formation of the nanomask, the thin GaAs buffer layer is grown. A hollow structure is formed on the surface of the buffer

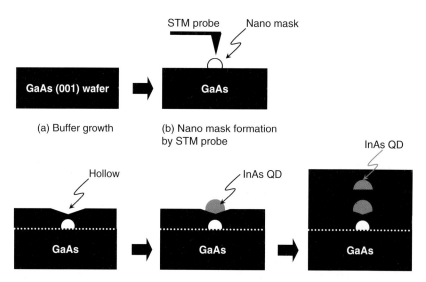

Figure 4.35 STM-probe-assisted nanolithography for fabrication of site-controlled QD.

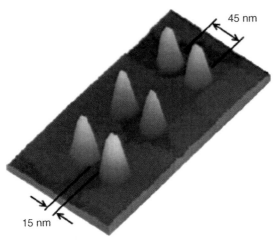

Figure 4.36 STM image of a site-controlled InAs QD structure on GaAs surface [35].

layer, as GaAs does not grow on the nanomask. A site-controlled QD structure can be fabricated from the selective growth of the InGaAs QD on this hollow structure. Figure 4.36 shows a typical STM image of the site-controlled InAs QD structure on GaAs using the STM-probe assisted nanolithography technique. It is observed that the site-controlling resolution is approximately of the order of nanometers. The height and dimensions of the site-controlled QD are similar to those of a self-assembled QD structure using a conventional fabrication technique. Therefore, the creation of quantum communication devices may be anticipated, since a quantum size effect is expected in the case of this QD.

4.3.5
Metal-Mask MBE Technique for Selective-Area QD Growth [36]

Regions of the self-assembled QD formation can be selected on the wafer by using a metal mask (MM) MBE technique. It is well known that the typical cavity length of QD lasers is a few hundred micrometers. Therefore, a micrometer- or millimeter-sized area controlling of the QD formation constitutes an important technique for fabricating integrated photonic devices such as semiconductor lasers, optical amplifiers, etc. This chapter introduces the selective-area growth technique for QDs using a metal mask. In the MBE method, a molecular flux is irradiated onto a substrate surface for crystal growth. Therefore, an area of the QD formation is selected, while the molecular-flux intensity is controlled using a mask near the substrate surface. Figure 4.37 shows the setup for the selective-area growth technique for QD formation using a metal mask. There are windows a few hundred micrometers in size on the mask, and the mask is closed over the substrate surface. A self-assembled QD structure can be obtained only in the area of the windows; it is also confirmed that there is no QD growth on the masked regions.

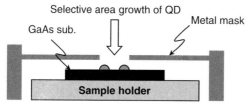

Figure 4.37 Cross-sectional image of the selective-area growth technique of MBE using a metal mask.

4.4
Silicon-Related Quantum Structure Fabrication Technology

This chapter introduces the successful fabrication of a III-V compound semiconductor QD structure on a silicon substrate. Additionally, a fabrication technique for silicon nanoparticles and related materials, such as silicon quantum dots, is also introduced.

4.4.1
III-V Compound Semiconductor QD on a Si Substrate [37,38]

Silicon is a widely used substrate in electronic integrated circuits. The fusion of conventional electronic integrated circuits and nanophotonics is anticipated to emerge as a breakthrough technology in the future. A III–V compound semiconductor QD structure fabricated on a Si substrate is therefore a likely candidate for developing this technology.

This section introduces a fabrication technique for the III–V compound semiconductor QD on a Si substrate. In particular, the formation of a III-Sb compound semiconductor on Si is explained. Figure 4.38 shows a typical AFM image of a fabricated Sb-based QD on a Si substrate. This QD structure can be obtained using a conventional MBE method. The fabrication sequence of the Sb-based QD on the Si substrate is as follows. *I.* An Si (001) substrate surface is prepared by chemical etching using an HF solution, and a nature oxide layer is removed; thereafter, the surface is terminated with hydrogen. *II.* The Si surface is also prepared by thermal cleaning at 800 °C in the MBE chamber. *III.* The substrate temperature is lowered to 400 °C before the QD formation. *IV.* Molecular fluxes of In, Ga, and Sb are irradiated simultaneously onto the Si substrate in order to grow the Sb-based QD structure; thereafter, an approximately 2 ML (0.1 ML/s) layer is deposited. *V.* Sb-irradiation is carried out a few times after the QD formation. At this point, the Sb-based QD structure is influenced with an Sb-flux intensity, as shown in Figure 4.39. Therefore, optimization of the Sb-flux is important for obtaining a high-quality Sb-based QD on Si. A nanopositioning technique with a patterned SiO_2 mask, as explained in Section 4.3.1, can be effectively carried out, since the SiO_2 film is widely used in the processes of Si devices. Nanostructured Sb-based semiconductors, such as Sb-based QD structures, are expected to emerge as useful materials for the development of an Si-MOS compatible nanophotonics technology.

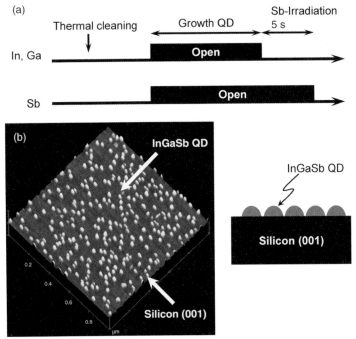

Figure 4.38 (a) Shatter sequence of MBE for growth of Sb-based QD on Si, (b) typical AFM image of the fabricated Sb-based QD structure on Si (001).

4.4.2
Fabrication Technique of Silicon Nanoparticles as Si-QD Structures [39]

Compound semiconductors are generally widely studied for the fabrication of the QD structure because they result in an observable quantum size effect in the quantum-confinement structure of a relatively large size (approximately a few tens

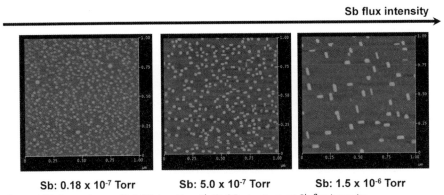

Figure 4.39 Dependence of 1.5-ML $In_{0.5}Ga_{0.5}Sb$/Si QD structure on Sb-flux intensity.

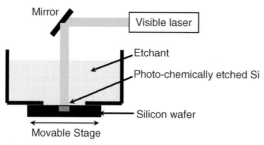

Figure 4.40 Photochemical etching method for fabrication of Si nanoparticles.

of nanometers). On the other hand, a carrier-confined structure several nanometers in size, which is generally called a nanoparticle, is necessary when using a silicon semiconductor material. Several techniques have been proposed for the fabrication of the Si nanoparticle as an Si-QD structure. An anodization method and a photochemical etching method of a Si wafer are proposed for producing the Si nanoparticle in a simple manner [39–41]. It is known that the Si nanoparticle exhibits a bright visible light emission of red or blue color, and it is considered that this light emission is caused by the quantum size effect of the Si-QD. In addition, electroluminescence devices on an Si wafer are also demonstrated using Si nanoparticles [42]. In particular, this section introduces the photochemical etching method for the fabrication of Si nanoparticles. Figure 4.40 shows a schematic setup diagram of the photochemical etching method. An n-type Si wafer (001) is set at the bottom of a vessel filled with an etchant. A mixture of hydrofluoric acid solution (HF)

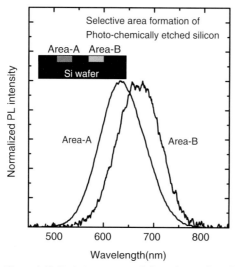

Figure 4.41 Emission spectra of photochemically etched layers. The emission colors are red and yellow. Each layer is formed on the same Si substrate using a selective-area formation technique.

and hydrogen peroxide (H_2O_2) as an oxidant is used as the etchant, with the typical concentration of the etchant being $HF:H_2O_2 = 100:17$. The Si wafer surface is irradiated by a visible laser (for example, He-Ne laser) for a few tens of minutes in order to form a photochemically etched layer, which also emits a visible photoluminescence. In practice, it is possible to realize a selective-area formation of the photochemically etched layer, such as Si nanoparticles, by controlling the position of the laser irradiation shown in Figure 4.41. Additionally, it is expected that selective-area formation of the Si-QD structure of submicrometer scale will be realized using near-field optics for the photochemical etching.

References

1 Arakawa, Y. and Sakaki, H. (1982) *Applied Physics Letters*, **40**, 939–941.
2 Miyazawa, Toshiyuki, Takemoto, Kazuya, Sakuma, Yoshiki, Hirose, Shinichi, Usuki, Tatsuya, Yokoyama, Naoki, Takatsu, Motomu and Arakawa, Yasuhiko (2005) *Japanese Journal of Applied Physics*, **44**, L620–L622.
3 Ohtsu, Motoichi, Kobayashi, Kiyoshi, Kawazoe, Tadashi, Sangu, Suguru and Yatsui, Takashi (2002) *IEEE Journal of Selected Topics in Quantum Electronics*, **8**, 839–862.
4 Goldstein, L., Glas, F., Marzin, J.Y., Charasse, M.N. and Le Roux, G. (1985) *Applied Physics Letters*, **47**, 1099–1101.
5 Stranski, I.N. and Krastanow, L. (1937) *Akad Wiss Wien Math –Natur IIb*, **146**, 797.
6 Huang, X.D., Stintz, A., Hains, C.P., Liu, G.T., Cheng, J. and Malloy, K.J. (2000) *IEEE Photonics Technology Letters*, **12**, 227.
7 Markus, Chen, J.X., Paranthoen, C., Fiore, A., Platz, C. and Gauthier-Lafaye O. (2003) *Applied Physics Letters*, **82**, 1818.
8 Wang, J.S., Hsiao, R.S., Chen, J.F., Yang, C.S., Lin, G., Liang, C.Y., Lai, C.M., Liu, H.Y., Chi, T.W. and Chi, J.Y. (2005) *IEEE Photonics Technology Letters*, **17**, 1590.
9 Liu, C.Y., Yoon, S.F., Cao, Q., Tong, C.Z. and Li, H.F. (2007) *Applied Physics Letters*, **90**, 041103.
10 Kovsh, R., Maleev, N.A., Zhukov, A.E., Mikhrin, S.S., Vasil'ev, A.P., Shernyakov, Yu.M., Maximov, M.V., Livshits, D.A., Ustinov, V.M., Alferov, Zh.I., Ledentsof, N.N. and Bimberg, D. (2002) *Electronics Letters*, **38**, 1104.
11 Ishida, M., Hatori, N., Akiyama, T., Otsubo, K., Nakata, Y., Ebe, H., Sugawara, M. and Arakawa, Y. (2004) *Applied Physics Letters*, **85**, 4145.
12 Sugawara, M., Hatori, N., Ishida, M., Ebe, H., Arakawa, Y., Akiyama, T., Otsubo, K., Yamamoto, T. and Nakata, Y. (2005) *Journal of Physics D-Applied Physics*, **38**, 2126.
13 Saito, H., Nishi, K., Kamei, A. and Sugou, S. (2000) *IEEE Photonics Technology Letters*, **12**, 1298.
14 Ohtsu M. (ed.) (1998) *Near-Field Nano/Atom Optics and Technology*, Springer, Tokyo.
15 Kawazoe, T., Kobayashi, K., Akahane, K., Naruse, M., Yamamoto, N. and Ohtsu, M. (2006) *Applied Physics B-Lasers and Optics*, **84**, 243.
16 Kaspi, R., Reynolds, D.C., Evans, K.R. and Taylor, E.N. (1994) *Institute of Physics Conference Series*, **141**, 57.
17 Shimizu, H., Kumada, K., Uchiyama, S. and Kasukawa, A. (2000) *Electronics Letters*, **36**, 1379.
18 Harmand, J.C., Li, L.H., Patriarche, G. and Travers, L. (2004) *Applied Physics Letters*, **84**, 3981.
19 Larsson, Mats I., Ni, Wei.-Xin, Joelsson, Kenneth and Hansson, Goran V. (1994) *Applied Physics Letters*, **65**, 1409.

20 Oh, Chan Wuk, Lee, Young Hee and Lee, Hyung Jae (1995) *Institute of Physics Conference Series*, **145**, 1267.

21 Portavoce, A., Berbezier, I. and Ronda, A. (2003) *Materials and Engineering B*, **101**, 181.

22 Tanaka, Satoru, Takeuchi, Misaichi and Aoyagi, Yoshinobu (2000) *Japanese Journal of Applied Physics*, **39**, L831.

23 Yamamoto, Naokatsu, Akahane, Kouichi, Gozu, Shinichirou and Ohtani, Naoki (2005) *Applied Physics Letters*, **86**, 203118.

24 Yamamoto, Naokatsu, Akahane, Kouichi, Gozu, Shin-ichirou, Ueta, Akio and Ohtani, Naoki (2006) *Japanese Journal of Applied Physics*, **45**, 3423.

25 Akahane, Kouichi, Yamamoto, Naokatsu, Ohtani, Naoki, Okada, Yoshitaka and Mitsuo, Kawabe (2003) *Journal of Crystal Growth*, **256**, 7.

26 Akahane, Kouichi, Kawamura, Takahiro, Okino, Kenji, Koyama, Hiromichi, Lan, Shen, Okada, Yoshitaka and Kawabe, Mitsuo (1998) *Applied Physics Letters*, **73**, 3411.

27 Springholz, G., Pinczolits, M., Holy, V., Zerlauth, S., Vavra, I. and Bauer, G. (2001) *Physica E*, **9**, 149–163.

28 Yamamoto, Naokatsu, Gozu, Shin-ichirou, Akahane, Kouichi, Ueta, Akio and Tsuchiya, Masahiro (2007) 28th International Conference on the Physics of Semiconductors – ICPS 2006. AIP Conference Proceedings, Volume 893, pp. 69–70.

29 Baier, M.H., Pelucchi, E., Kapon, E., Varoutsis, S., Gallart, M., Robert-Philip, I. and Abram, I. (2004) *Applied Physics Letters*, **84**, 648.

30 Iwamoto, Satoshi, Jun, Tatebayashi, Tatsuya, Fukuda, Toshihiro, Nakaoka, Satomi, Ishida and Yasuhiko, Arakawa (2005) *Japanese Journal of Applied Physics*, **44**, 2579–2583.

31 Yamamoto, Naokatsu, Akahane, Kouichi, Gozu, Shinichiro and Ohtani, Naoki (2005) *Applied Physics Letters*, **86**, 203118.

32 Yamamoto, Naokatsu, Akahane, Kouichi, Gozu, Shin-ichirou, Ueta, Akio and Ohtani, Naoki (2006) *Japanese Journal of Applied Physics*, **45**, 3423.

33 Nakamura, Yusui, Ikeda, Naoki, Ohkouchi, Shunsuke, Sugimoto, Yoshimasa., Nakamura, Hitoshi and Asakawa, Kiyoshi (2004) *Japanese Journal of Applied Physics*, **43** (3A), L362–364.

34 Ohkouchi, S., Sugimoto, Y., Ozaki, N., Ishikawa, H. and Asakawa, K. (2008) *Physica E*, **40**, 1794–1796.

35 Kohmoto, Shigeru, Nakamura, Hitoshi, Ishikawa, Tomonori and Asakawa, Kiyoshi (1999) *Applied Physics Letters*, **75** (22), 3488.

36 Ozaki, N., Takata, Y., Ohkouchi, S., Sugimoto, Y., Nakamura, Y., Ikeda, N. and Asakawa, K. (2007) *Journal of Crystal Growth*, **301–302**, 771–775.

37 Yamamoto, Naokatsu, Akahane, Kouichi, Gozu, Shin-ichirou, Ueta, Akio, Ohtani, Naoki and Tsuchiya, Masahiro (2007) *Japanese Journal of Applied Physics*, **46** (4B), 2401–2404.

38 Yamamoto, N., Akahane, K., Gozu, S., Ueta, A. and Tsuchiya, M. (2008) *Physica E*, **40** (6), 2195–2197.

39 Yamamoto, Naokatsu and Takai, Hiroshi (1999) *Japanese Journal of Applied Physics*, **38**, 5706–5709, or *Thin Solid Films* **388** (2001) pp. 138–142.

40 Canham, L.T. (1990) *Applied Physics Letters*, **57**, 1046.

41 Hadjersi, T., Gabouze, N., Yamamoto, N., Sakamaki, K. and Takai, H. (2004) *Thin Solid Films*, **459**, 249–253.

42 Yamamoto, Naokatsu, Sumiya, Atsushi and Takai, Hiroshi (2000) *Materials Science & Engineering*, **B69–70**, 205–209.

5
ZnO Nanorod Heterostructures for Nanophotonic Device Applications
Gyu-Chul Yi

5.1
Introduction

One-dimensional (1D) semiconductor nanostructures, including nanorods, nanowires, nanobelts, and nanotubes, are vital components for fabricating optoelectronic and photonic nanodevices in bottom-up approaches [1–3]. Recently, accurate controls of both compositions and layer thicknesses have produced nanorod heterostructures exhibiting the quantum-confinement effect [4–6]. Embedding quantum structures in a single nanorod would enable exploration of novel physical properties such as quantum confinement and continuous tuning of the spectral wavelength by varying the well thickness. Since an important component for fabricating nanophotonic devices and understanding their operation is dealing with light–matter interaction at the nanometer scale, quantum effects in nanorod quantum structures comprise the key feature in nanophotonic device operations [3, 7, 8].

As illustrated in Figure 5.1(a), two basic types of nanorod quantum structures exist: axial nanorod quantum well (QW) structures and radial (coaxial) nanorod heterostructures. The former exhibit composition modulation along the axial direction, where the heterointerfaces between the well and barrier layers are perpendicular to the rod axis, and the latter composition modulation along the radial direction with the heterointerfaces parallel to the rod axis. These nanorod quantum structures can be used for fabricating sophisticated nanometer-scale photonic, electronic, and optoelectronic devices (Figure 5.1(b)). For the axial structures, both carriers and excitons can be confined in the QW layer along the axial direction of a nanorod, which is useful for optoelectronic and photonic devices such as nanophotonic switches [3] and light-emitting diodes [9–13], as well as nanoelectronics for resonant tunneling devices [14] and single-electron transistors [15]. Radial nanorod quantum structures having well-defined interfaces are employed as the principal components in nanoscale light-emitting devices [8] and high-speed electronic devices [16]. The radial nanorod quantum structures can also play important roles as both optical interconnects and as functional units in fabricating optoelectronic and photonic

Nanophotonics and Nanofabrication. Edited by Motoichi Ohtsu
Copyright © 2009 WILEY-VCH Verlag GmbH & Co. KGaA, Weinheim
ISBN: 978-3-527-32121-6

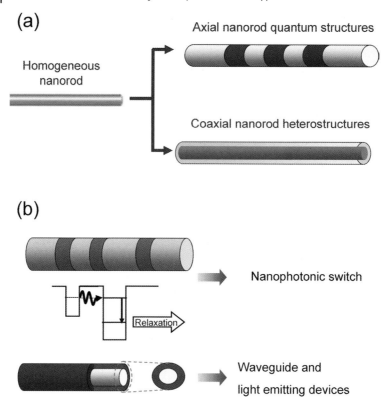

Figure 5.1 (a) Schematics of two types of nanorod heterostructures showing the composition modulation along the axial and radial directions of a nanorod. (b) Possible nanoelectronic and nanophotonic device applications of nanorod quantum structures.

nanodevices [17–19]. Virtues of the nanorod quantum structures include not only the miniaturization of nanophotonic devices, but also the presentation of a platform to investigate novel principles of functional operations inherent to nanophotonics. In particular, three-dimensional (3D) nanoarchitectures combining both axial and radial nanorod heterostructures offer an ideal component for nanophotonic device applications.

Because of their unique physical properties, ZnO semiconductor nanorods and their heterostructures are good candidates for nanoscale photonic device applications [20–22]. ZnO has a fundamental direct bandgap with a bandgap energy (E_g) of 3.4 eV at room temperature, and with an exciton binding energy as large as 60 meV, free excitons can survive even at room temperature. Furthermore, both n- and p-type ZnO nanorods [23–25] can be prepared although reliable and reproducible fabrication of light-emitting nanodevices based on p–n junction ZnO nanorods has not yet been reported. The good thermal and mechanical stabilities of ZnO nanorods offer further advantages [26, 27]. These optical, electrical, and mechanical characteristics of ZnO nanorods will be very useful in numerous photonic and optoelectronic device applications.

In the past several years, much effort has been devoted to developing various semiconductor nanorod heterostructures for nanodevice applications. Both metal catalyst-assisted vapor-liquid-solid (VLS) [14, 28, 29] and catalyst-free methods [30–34] have been widely used to synthesize various semiconductor nanorod heterostructures. In particular, catalyst-free metalorganic vapor-phase epitaxy (MOVPE) techniques have been employed in growing vertically aligned ZnO nanorods on many substrates including sapphire [30] and Si [35].

The catalyst-free MOVPE technique may have several advantages over the catalyst-assisted methods. First, the catalyst-free method enables growth of high-purity and single-crystalline semiconductor nanorods [35, 36]. Second, the computer-controlled reactant-gas delivery system of a MOVPE system makes thickness control at a monolayer level readily achievable: this is a great advantage in growing composition-modulated nanorod heterostructures with sharply defined interfaces [4, 6, 37]. In nanorod heterostructures of wide-bandgap materials, such as ZnO and GaN, the QW layer thickness must be smaller than a few nanometers for effective quantum confinement of charge carriers, which can be readily achieved by the MOVPE system [4]. Third, the catalyst-free MOVPE enables the growth of vertically aligned ZnO nanorods at temperatures below 400 °C. This low-temperature process is compatible with current microfabrication processes for conventional Si-based electronics, beneficial to monolithic integration of both electronic and photonic devices. The low-temperature growth of single-crystalline semiconductor nanorods also allows us to use various substrates including Al_2O_3 [30], Si [35], GaN, [37] glasses, polymers and metals [38] for the nanorod growth. Furthermore, recent reports on the position-controlled, selective growth of ZnO nanorod arrays on Si substrates [39, 40] demonstrate an important breakthrough for functional integration of nanophotonic and conventional electronic devices.

Clean interfaces between QW and barrier layers are required to prepare high-quality nanorod quantum structures. In particular, the formation of lattice defects, including misfit dislocations at the interface in the nanorod heterostructures, must be suppressed. These goals have been met in thin films by heteroepitaxially growing the QW layers in thin-film structures using MOVPE and molecular beam epitaxy (MBE). As mentioned above, semiconductor nanorod quantum structures, which are promising nanostructures for realizing nanophotonic devices, have been successfully prepared using catalyst-free MOVPE and hydride vapor-phase epitaxy [4, 6, 10, 41, 42], thereby enabling the exploration of quantum confinement in artificially fabricated nanorod heterostructures.

This chapter reviews recent research activities related to ZnO nanorods and their heterostructures, including nanorod quantum structures for nanoscale photonic device applications. The main text focuses on axial and radial nanorod heterostructures as classified by their composition modulation direction. Section 5.2 describes $ZnO/Zn_{0.8}Mg_{0.2}O$ axial nanorod heterostructures with single quantum-well structures (SQWs), multiple-quantum-well structures (MQWs), and quantum dots (QDs) in the MQWs and their optical characteristics. Section 5.3 discusses ZnO-based radial nanorod heterostructures having radial composition modulation and their structural and optical properties. The chapter concludes with personal remarks on the outlook for ZnO nanorod nanophotonic device applications.

5.2
ZnO Axial Nanorod Quantum Structures

This section describes fabrications of the ZnO/Zn$_{0.8}$Mg$_{0.2}$O axial nanorod quantum structures and their structural and optical properties. For the fabrications of nanorod quantum structures, the catalyst-free MOVPE among many semiconductor nanorod growth methods has been employed in order to control thicknesses and compositions of layers precisely. MOVPE, a well-developed but sophisticated thin-film technique used for many device applications, offers high-quality nanorod heterostructures as well as thin-film quantum structures. With precise thickness control down to the monolayer level, these heterostructures show the clear signature of quantum confinement. The tunability of layer thicknesses and compositions in the nanorod QW heterostructures greatly increases their versatility and power for device fabrications in nanoscale photonics.

The approach to fabricate high-quality nanorod heterostructures is to utilize catalyst-free MOVPE that does not need metal catalysts usually required in other catalyst-assisted methods. Especially, in addition to growth of homogeneous ZnO nanorods [30], vertically aligned ZnO/Zn$_{1-x}$Mg$_x$O ($x<0.3$) nanorod quantum structures can be easily grown using catalyst-free MOVPE [4, 5]. For the growths of Zn$_{1-x}$Mg$_x$O layers, *bis*-cyclopentadienyl-Mg (cp_2Mg) was used as the Mg precursor, with composition controlled through the partial pressure [43]. ZnO/Zn$_{1-x}$Mg$_x$O nanorod single and multiple QW structures were fabricated by controlling the reactant gas flow directions to either a reactor or vent. The QW and quantum barrier (QB) layer thicknesses were precisely controlled using a computer-controlled gas valve system in the MOVPE system [4]. Between each change in composition of the reaction gases, the reactor was purged with pure argon and oxygen, so no layer of mixed composition is expected to be generated. Abrupt interfaces between ZnO QW and Zn$_{1-x}$Mg$_x$O QB layers were achieved by direct adsorption of atoms on the top surface of nanorods in the catalyst-free MOVPE reactor.

At first, we fabricated ZnO/Zn$_{1-x}$Mg$_x$O axial MQWs and investigated their physical properties. As a schematic diagram of the nanorod MQWs is shown in Figure 5.2(a), the nanorod MQWs prepared in this study include 10 periods of ZnO/Zn$_{1-x}$Mg$_x$O QW structures, which were prepared only on tips of ZnO nanorods. The Mg content was chosen to be 20 at.% since Zn$_{1-x}$Mg$_x$O for $x=0.2$ can be a good QB to confine carriers in a ZnO well layer. In addition, small lattice constant and thermal expansion coefficient mismatches between the ZnO and Zn$_{0.8}$Mg$_{0.2}$O layers minimize formation of interfacial defects in the nanorod quantum structures [43]. For Zn$_{0.8}$Mg$_{0.2}$O/ZnO nanorod MQWs, we expect to observe quantum-confinement effect confining carriers in ZnO QW layers. Our simple theoretical calculations using the finite periodic square-well potential also predicted the quantum-confinement effect. The

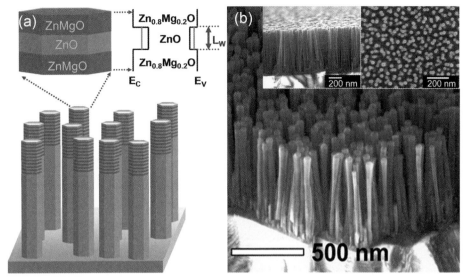

Figure 5.2 (a) Schematic of nanorod multiple QW structures (MQWs) including 10 periods of $Zn_{0.8}Mg_{0.2}O/ZnO$ QW structures on the tips of ZnO nanorods. An electronic band diagram for the MQW is also shown. (b) FEG-SEM images of the nanorod MQWs.

parameters used for this calculation are $0.28m_0$ and $1.8m_0$ for the effective masses of electron and hole in ZnO, respectively, the ratio of conduction and valence band offsets ($\Delta E_c/\Delta E_v$) of 9, and the bandgap offset (ΔE_g) of 250 meV [44].

Field-emission gun scanning electron microscopy (FEG-SEM) reveals general morphology of the nanorod axial MQWs. As shown in Figure 5.2(b), the nanorods were well aligned vertically with homogeneous length and diameter distributions. Typical mean diameters and lengths were in the range of 20–70 nm and 0.5–2 μm, respectively, and a normalized standard deviation value in length distributions was as small as 0.02, indicating that the nanorod MQWs exhibit uniform length distribution, and also implying uniform ZnO well-width distribution. Additionally, X-ray diffraction (XRD) results showed that the nanorod heterostructures were epitaxially grown with homogeneous in-plane alignment as well as c-axis orientation normal to Al_2O_3 (0001) substrate surface plane.

The detailed crystal structure of the nanorod MQWs was investigated using transmission electron microscopy (TEM). As shown in Figures 5.3(a) and (b), the interfaces between the different layers are clearly visible and TEM images show QW layer thicknesses of 1.1 and 2.5 nm, respectively; both images exhibit the bright and dark layers in the MQW, corresponding to the $Zn_{0.8}Mg_{0.2}O$ and ZnO layers, respectively. Figure 5.3(c) presents the high-resolution TEM image of the sample with a well-layer thickness of 2.5 nm, indicating that nanorod MQWs are grown along the c-axis orientation of the hexagonal crystal structure. The interfaces between ZnO and $Zn_{0.8}Mg_{0.2}O$ are as clean as lattice image in the ZnO layer, and

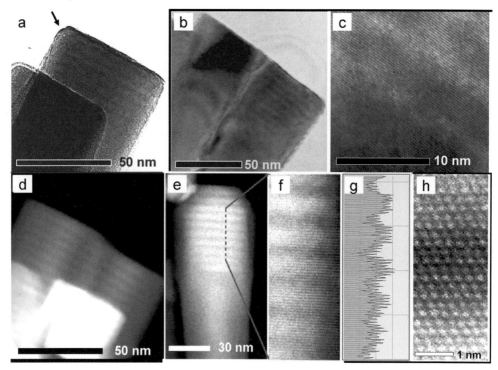

Figure 5.3 TEM images of ZnO/Zn$_{0.8}$Mg$_{0.2}$O nanorod MQWs. (a) and (b) Low-magnification TEM images of the samples with QW layer thicknesses of 1.1 and 2.5 nm. The repeating dark and bright layers represent the MQW region. (c) High-resolution TEM image of the 2.5 nm sample. (d)–(h) Z-contrast images of the 2.5 nm sample with increasing magnification and an intensity profile (g) along the dashed line in (e). The Z-contrast image clearly shows the compositional variation, with the bright layers representing the ZnO QW layers.

interface dislocations were rarely observed. The clean interface without any defect formation results presumably from the small lattice mismatches between ZnO and Zn$_{0.8}$Mg$_{0.2}$O and no memory effect in the MOVPE system [43].

Figures 5.3(d)–(h) show Z-contrast TEM images of the nanorod MQWs. As shown in Figure 5.3(f), the higher magnification image clearly shows the well-resolved ZnO lattices. The MQW period was determined to be 5.5 ± 0.1 nm. However, the intensity profile across this image (Figure 5.3(g)) shows that the boundaries between ZnO and Zn$_{0.8}$Mg$_{0.2}$O are not very clear. The apparent broadening of the interface is attributed to the slightly rounded shape of the nanorod growth surface near the sides of the rod (arrowed in Figure 5.3(a)). In projection, this will lead to an apparent blurring of the interface of the order of 1 nm, as observed. Meanwhile, no layer of mixed composition is expected as mentioned previously, and interdiffusion during the low-temperature growth at 450 °C is also negligible since ZnO/ZnMgO heterostructures are thermally stable up to 750 °C [45].

Quantum phenomena of ZnO/Zn$_{0.8}$Mg$_{0.2}$O nanorod MQWs were investigated by measuring photoluminescence (PL) spectroscopy at 10 K. As shown in Figure 5.4,

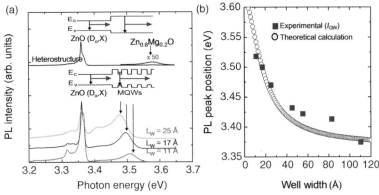

Figure 5.4 (a) 10-K PL spectra of ZnO/Zn$_{0.8}$Mg$_{0.2}$O nanorod heterostructures and MQWs with band diagrams shown in the inset. (b) Well-width-dependent PL peak positions in ZnO/Zn$_{0.8}$Mg$_{0.2}$O nanorod MQWs (squares) and theoretically calculated values (circles) in 10 periods of one-dimensional square potential wells.

the PL spectra of the nanorod MQWs exhibit a common PL peak at 3.360 eV and an additional PL peak at 3.478, 3.496 and 3.515 eV for QW widths of 2.5, 1.7 and 1.1 nm, respectively. Compared with the PL peaks from ZnO/Zn$_{0.8}$Mg$_{0.2}$O nanorod heterostructures and homogeneous ZnO nanorods and thin films [36, 46, 47], the PL peak (I^{ZnO}) at 3.360 eV is attributed to the neutral-donor bound excitons in ZnO, resulting from ZnO nanorod stems. Meanwhile, the PL peak (I^{ZnMgO}) at 3.58 eV observed from the Zn$_{0.8}$Mg$_{0.2}$O in the nanorod heterostructures was not observed for the nanorod MQWs presumably because carriers are effectively confined into the ZnO QW layers.

Variation in the I_{QW} peak energy depending on the ZnO well-layer width as well as theoretical predictions in finite square-well potential are depicted in Figure 5.4(b). Results from other samples are also included in Figure 5.4(b); the blue-shift decreases with increasing well width and is almost negligible at a well width of 110 Å. The simple theoretical predictions using the finite periodic square-well potential show that the systematic increase in PL emission energy with reducing well width results from the quantum-confinement effect.

The PL spectrum of ZnO/Zn$_{0.8}$Mg$_{0.2}$O nanorod MQWs with a well-layer thickness of 2.5 nm, as shown in Figure 5.4(a), exhibits an additional PL peak at 3.41 eV between I^{ZnO} and I_{QW}. To investigate the origin of the PL peak, TEM and PL spectra of ZnO/Zn$_{0.8}$Mg$_{0.2}$O nanorod MQWs with various QW layer thicknesses were further measured [48]. Figures 5.5 (a) and (b) present the TEM images of ZnO/Zn$_{0.8}$Mg$_{0.2}$O nanorod MQWs with different ZnO QW layer thickness (L_w) of 11 Å and 44 Å, revealing the dark and bright layers corresponding to ZnO and Zn$_{0.8}$Mg$_{0.2}$O layers, respectively. The distinctive feature of the overall morphology of ZnO well layers, depending on the QW layer thickness was observed. That is, the sample with 11 Å has uniformly thick disk-like layers [Figure 5.5(a)], while the sample with 44 Å exhibits the truncated pyramid shape on a disk-like plate [Figure 5.5(b)]. The TEM images offer the clear evidence on the formation of QDs in the top part of the nanorod.

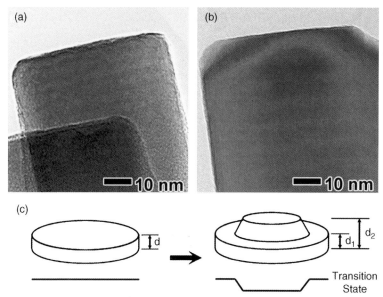

Figure 5.5 (a) TEM image of ZnO/ZnMgO nanorod MQWs with $L_w = 11$ Å. (b) TEM image of ZnO/ZnMgO nanorod MQWs with $L_w = 44$ Å. Note that a quantum-dot-like structure is only observed in the sample with $L_w = 44$ Å. (c) Schematic diagrams of ZnO QW layer shape across the nanorod and relative height of transition energy.

The growth mode transition from quasi-2-dimensional (2D) plates to 3D island or QD structures occurred between 11 Å and 44 Å thickness of the QW layers.

Schematic diagrams of two different well-layer structures and related interband transition energy are illustrated in Figure 5.5(c). In the case of MQWs with uniformly thin ZnO well layers, a unique interband transition state exists. On the other hand, the excitons localized at the fringes of the well layers have the higher transition energy due to narrower thickness of d_1, while those at the truncated pyramids with the wider thickness of d_2 exhibit the lower transition energy. Thus, the band-edge emission consists of two components; one from the excitonic contribution in MQWs and the other from QD structure.

Apparently, from TEM and SEM measurements, the nanorod QD structures can affect radiative transition, so that the PL spectra of ZnO/Zn$_{0.8}$Mg$_{0.2}$O nanorod MQWs show different behavior depending on the MQW widths. From 10-K PL spectra of ZnO/Zn$_{0.8}$Mg$_{0.2}$O nanorod MQWs with various well widths (L_w), as shown in Figure 5.6, the additional peak (I_{QD}) starts to appear at 3.411 eV for the nanorod MQWs with $L_w = 25$ Å, and was observed for all nanorod MQWs with $L_w \geq 25$ Å. As the well width was increased further, the PL emission energies of I_{QW} and I_{QD} decrease slightly. The overall red-shift of I_{QD} peak position results from decreased quantum-confinement effect with increasing L_w. The similar behavior of I_{QW} peak position was also observed, and can be explained with the same argument.

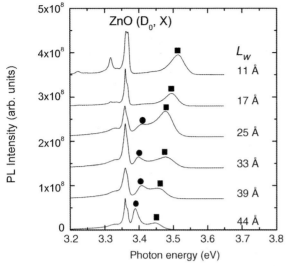

Figure 5.6 Low-temperature (measured at 10 K) PL spectra of ZnO/Zn$_{0.8}$Mg$_{0.2}$O MQW structured nanorods with different well widths. Well-width-dependent peak positions of emission peaks marked by I_{QW} (■) and I_{QD} (●).

In addition to the appearance of the new PL peak of I_{QD}, Figure 5.6 also shows that the relative ratio of integrated intensity of I_{QD} over peak I_{QW} increased drastically from 0.34 for a L_w of 25 Å to 3.1 for a L_w of 44 Å. For the nanorod MQWs with thicker well layers, more QDs are formed. In addition, large-size QDs have lower energy states than thin QWs do, confining more carriers in QDs. Accordingly, as the well-layer thickness increases, the luminescence from QD becomes more dominant than that of QWs.

As shown in Figure 5.5(b), it seems that the QDs were formed only in the QW layers near the nanorod tips, so that the QD formation depending the QW period was further investigated. Figure 5.7 shows the SEM image and low-temperature PL spectra for ZnO/Zn$_{0.8}$Mg$_{0.2}$O nanorod SQWs and MQWs with 4, 7 and 10 periods. The SEM images clearly show that the tip morphology is flat for the nanorod SQWs, but not for the nanorod MQWs with ten periods. From the TEM images in Figure 5.5, this nonflat SEM morphology can be directly associated with QD formation. In addition, the PL spectra show that the PL peak at 3.42 eV, which is attributed to radiative transition in QD, starts to appear from the sample with the 4 periods. From the PL spectra and SEM images, it can be inferred that QD formation does not take place for a nanorod single QW since the strain field for a single QW was not large enough to create a QD. The selective formation of QD only in MQWs may result from strain field accumulated from each QW layer as the strain-induced QD formation is well known for thin film-structures [49].

The QD formation in ZnO/ZnMgO nanorods can be understood in terms of strain relaxation. That is, the unrelieved strain field propagates through the growth direction of nanorods and can cause QD formation when the accumulated strain

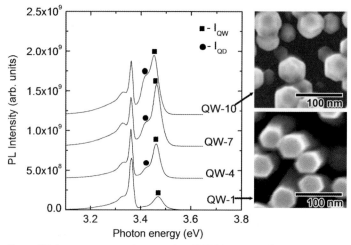

Figure 5.7 Low-temperature (measured at 10 K) PL spectra of ZnO/Zn$_{0.8}$Mg$_{0.2}$O MQW structured nanorods with different periods. SEM images show the morphology at the tip of nanorods, indicating the formation of QDs for the nanorod MQWs.

from each layer of MQWs structures is over a critical value. Since previous reports argued that unlike thin films, the strain due to lattice mismatch in nanorods was relieved within a few atomic layers from the interface [50], it is very interesting if the strain is not fully relaxed so that QDs are formed.

In addition to the nanorod MQWs, ZnO/Zn$_{0.8}$Mg$_{0.2}$O nanorod single QWs (SQWs) were fabricated as the schematic is shown in Figure 5.8(a). ZnO well-layer thickness (L_w) investigated in this study ranged from 11 to 90 Å while the thicknesses of Zn$_{0.8}$Mg$_{0.2}$O bottom and top barrier layers in nanorod SQWs were fixed to 300 and 60 Å, respectively. Electron microscopy images show the general morphology of ZnO/Zn$_{0.8}$Mg$_{0.2}$O nanorod SQW arrays. As shown in Figure 5.8(b) and (c), well-aligned nanorods with a mean diameter of 40 nm were grown vertically. A typical mean length of ZnO/Zn$_{0.8}$Mg$_{0.2}$O nanorod SQWs was 970 ± 20 nm.

The detailed crystal structure of the nanorod SQWs was also investigated using TEM. Figure 5.8(d) presents a TEM image of the ZnO/Zn$_{0.8}$Mg$_{0.2}$O nanorod SQWs with a well width of 60 Å. Because of composition modulation along the individual nanorod tip ends, Z-contrast TEM images of nanorod SQWs exhibited the contrast change originated from the difference between electron scattering cross-sections. As shown in Figure 5.8(d), the Z-contrast image shows that a bright ZnO layer between Zn$_{0.8}$Mg$_{0.2}$O dark layers is 60 Å thick. The interface between the ZnO layer on Zn$_{0.8}$Mg$_{0.2}$O was very clean and abrupt (indicated by the solid line with an arrow), whereas the interface of Zn$_{0.8}$Mg$_{0.2}$O on ZnO (indicated by the dashed line with an arrow) is not so clearly defined, resulting presumably from interface roughness. Further HR-TEM measurements revealed that the interface roughness increases with ZnO layer thickness, resulting presumably from island growth mode of ZnO on Zn$_{0.8}$Mg$_{0.2}$O.

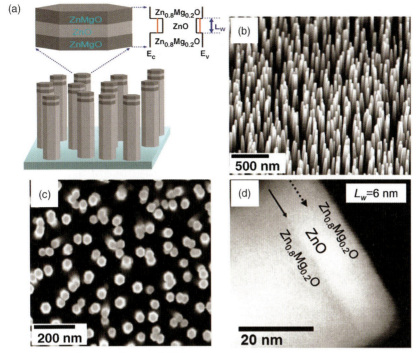

Figure 5.8 (a) Schematic of nanorod SQW structures consisting of $Zn_{0.8}Mg_{0.2}O/ZnO/Zn_{0.8}Mg_{0.2}O$ on the tips of ZnO nanorods and SQW electronic band diagram. (b) Tilted view and (c) plan-view FEG-SEM images of nanorod SQWs. (d) TEM image of nanorod SQWs.

Low-temperature PL spectra of $ZnO/Zn_{0.8}Mg_{0.2}O$ nanorod heterostructure (with a $Zn_{0.8}Mg_{0.2}O$ layer thickness of 200 nm) and SQWs with different ZnO well-layer thickness were measured in order to observe quantum-confinement effects in the nanorod SQWs. As shown in Figure 5.9(a), the nanorod heterostructure exhibited only PL peaks at 3.360 eV (I_5^{ZnO}) and 3.58 eV (I^{ZnMgO}) due to the light emissions from ZnO nanorod stems and ZnMgO layers, respectively. No peak between the ZnO and ZnMgO peaks was observed, implying no composition intermixing at the interfaces. Meanwhile, the nanorod SQWs exhibited PL peaks at 3.360–3.366 eV and 3.375–3.518 eV, and the PL peak energy blue-shifted from 10 meV to 160 meV as the well-layer width decreased from 90 to 11 Å. Variation in the I_{QW} peak energy depending on the QW width as well as theoretical predictions in finite square-well potential are depicted in Figure 5.9(c). The experimental data agrees well with the results from theoretical calculations, indicating that the systematic increase in the PL emission peak by reducing the well-layer width results from the quantum-confinement effect. Furthermore, as shown in Figure 5.9(b), electroreflectance (ER) spectra of the nanorod quantum structures at 90 K clearly confirms the existence of the quantized state.

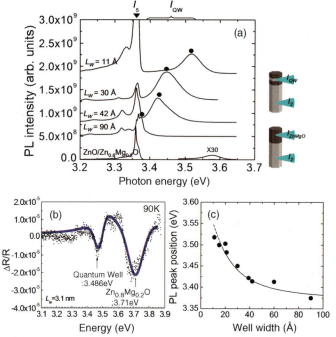

Figure 5.9 (a) 10-K PL spectra of ZnO/Zn$_{0.8}$Mg$_{0.2}$O nanorod heterostructure and SQWs with different ZnO well-layer widths, (b) 90-K ER spectra of nanorod SQWs with well width of 3.1 nm and (c) ZnO well-layer width vs. PL peak energy position in ZnO/Zn$_{0.8}$Mg$_{0.2}$O nanorod SQWs (closed circles) and theoretically calculated values (dashed curve).

It is remarkable that the nanorod SQWs showed excellent PL properties including strong room-temperature luminescence, even though the nanorod SQWs were grown on Si substrates with a large lattice mismatch with ZnO. Since the PL thermal quenching rate is generally dependent on interface quality, the low thermal quenching and strong PL at room temperature implies a high structural quality of SQWs with clean ZnO/ZnMgO interfaces. High-resolution TEM also confirmed that most defects caused by large lattice mismatches were formed at the interface of ZnO/substrate rather than inside nanorod SQWs [5], which leads to improved interface quality of the ZnO/ZnMgO QW layer on nanorods.

The utilization of QW structures restricts carrier motion to quasi-two dimensions and the confinement of carriers at the nanometer-scale gives rise to various quantum effects on optical characteristics although the optical characteristics of semiconductor bulk materials are mainly determined by the inherent band structure of the material. As the examples of the nanorod quantum structures, a series of ZnO/Zn$_{0.8}$Mg$_{0.2}$O nanorod with SQWs and MQWs have been fabricated on the top surfaces of the nanorods using catalyst-free MOVPE. Further reduction of dimensionality was achieved through QDs embedded in nanorod MQWs. The QD formation has the advantage of higher radiative recombination due to changed density of states, which results in decreasing the lasing threshold, improving thermal stability and exhibiting a narrow spectral lineshape.

5.3
ZnO Radial Nanorod Heterostructures

In this section, $ZnO/Zn_{0.8}Mg_{0.2}O$ radial nanorod heterostructures that exhibit the quantum-confinement effect are described. The radial nanorod heterostructures with composition modulations along a radial direction of a nanorod have a great potential as quantum building blocks for the fabrication of the nanometer-scale devices that utilize confinement of carriers, emitted light or both. For this nanorod heterostructures, in particular, nonradiative recombination by surface states can be reduced significantly by passivating surface states, resulting in enhancing the emitted light intensity at room temperature due to reduced thermal quenching of emission intensity.

$ZnO/Zn_{1-x}Mg_xO$ radial nanorod heterostructures were fabricated on Si substrates using catalyst-free MOVPE. The synthesis of core ZnO nanorods was carried out using DEZn and oxygen as the reactants with argon as the carrier gas. Depending on the growth conditions, the mean diameters of ZnO nanorods were in the range of 8–40 nm. The nanorod diameters were determined by TEM. Subsequent depositions of $Zn_{1-x}Mg_xO$ shell layer were performed by addition of cp_2Mg as the Mg precursor at the same chamber, resulting in ZnMgO layer coating on entire surfaces of ZnO nanorods. The shell layer coating on the ultrafine ZnO nanorods was robust and the ZnMgO layers were not so easily peeled off. The $Zn_{1-x}Mg_xO$ shell-layer thicknesses increased in proportion to the growth time with a typical growth rate of 3.2 nm/s. Average concentration of Mg in the $Zn_{1-x}Mg_xO$ layers was about 20 at.% as determined by energy-dispersive X-ray spectroscopy in the TEM chamber.

Synchrotron radiation (SR)-XRD was employed to investigate structural characteristics of ZnO nanorods and $ZnO/Zn_{0.8}Mg_{0.2}O$ radial nanorod heterostructures. From the SR-XRD $\theta-2\theta$ scan data of the ZnO nanorods, as shown in Figure 5.10, a dominant peak was observed at 29.04°, corresponding to ZnO(0002). However, a $Zn_{0.8}Mg_{0.2}O$ shell layer deposited on ZnO nanorods shifted the XRD peak to a higher angle with a slight increase in its full width at half-maximum value, depending on the $Zn_{0.8}Mg_{0.2}O$ layer thickness. For $ZnO/Zn_{0.8}Mg_{0.2}O$ radial nanorod heterostructures with a $Zn_{0.8}Mg_{0.2}O$ layer thickness of 4, 6, and 13 nm, furthermore, SR-XRD peak shifts to 29.05°, 29.06°, and 29.07°, respectively. The peak shift in the XRD curves originates from difference between ZnO and $Zn_{0.8}Mg_{0.2}O$ lattice constants. Due to their small lattice mismatch less than 0.5% [43], however, a thin $Zn_{0.8}Mg_{0.2}O$ layer can be heteroepitaxially grown on ZnO nanorods, yielding the only small peak shift without any signature of the peak separation in the resolution limit of SR-XRD measurements.

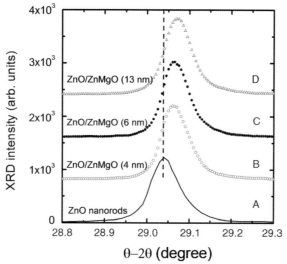

Figure 5.10 SR-XRD patterns of ultrafine ZnO and ZnO/ $Zn_{0.8}Mg_{0.2}O$ radial nanorod heterostructures with different shell-layer thicknesses of 4–13 nm. SR-XRD measurements were performed with the X-ray wavelength of 1.305 Å at the Pohang Accelerator Laboratory.

HR-TEM clearly reveals the lattice images of as-grown ZnO nanorods and their radial nanorod heterostructures, providing further structural analysis. In Figure 5.11(a), the diameter of as-grown ultrafine ZnO nanorods was as small as 8 nm. Additionally a highly ordered lattice image of the ultrafine ZnO nanorod is clearly displayed, indicating that ZnO nanorods are single crystalline without exhibiting any significant structural defects. Meanwhile, as shown in Figures 5.11(b)–(d), HR-TEM images of the radial heterostructure with a $Zn_{0.8}Mg_{0.2}O$ shell layer coated on a ZnO nanorod show the contrast change between core and shell layers, confirming the composition difference along the radial direction. Thickness of the $Zn_{0.8}Mg_{0.2}O$ shell layer was in the range of 11–14 nm. More importantly, the interfaces between the ZnO and $Zn_{0.8}Mg_{0.2}O$ layers were clean and abrupt as indicated by arrows in the TEM images. Dislocations were hardly observed in the nanorod heterostructures. These TEM results strongly suggest that the growth of $Zn_{0.8}Mg_{0.2}O$ on ZnO is coherently epitaxial, consistent with the XRD results.

The quantum-confinement effect along the radial direction in ultrafine ZnO nanorods was also investigated measuring PL spectroscopy. Figure 5.12(a) shows room-temperature PL spectra of ZnO nanorods with different mean diameters of 8, 9, 12, and 35 nm. While a dominant PL peak for the 35-nm thick ZnO nanorods was observed at 3.285 eV, the almost same position to that of bulk ZnO [46], the thinner nanorods show the blue-shift of the dominant PL peak; for ZnO nanorods with the diameters of 12, 9, and 8 nm, the PL peak was observed at 3.301, 3.316, and 3.327 eV with corresponding PL peak shifts of 16, 31, and 42 meV, respectively. The systematic blue-shift in their PL peak position with decreasing their diameter

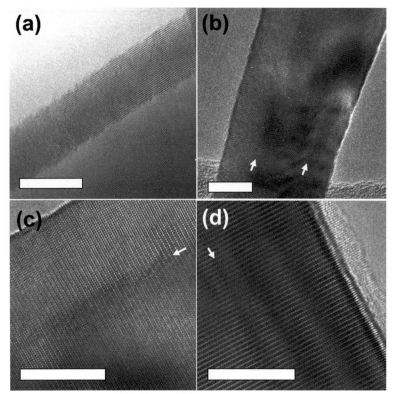

Figure 5.11 HR-TEM images of (a) ultrafine ZnO nanorods with a mean diameter of 8 nm and (b)–(d) ZnO/Zn$_{0.8}$Mg$_{0.2}$O radial nanorod heterostructures with a Zn$_{0.8}$Mg$_{0.2}$O shell layer. Arrows indicate the interfaces between ZnO core and Zn$_{0.8}$Mg$_{0.2}$O shell layers. The lengths of all bars correspond to 10 nm.

results presumably from the quantum-confinement effect along the radial direction in ZnO nanorods.

To clarify this quantum size effect, the energy shift (ΔE) was calculated using a cylindrical potential based on the effective mass model with a simple finite potential barrier. The ΔE was calculated to be 24, 44, and 55 meV for nanorod diameters of 12, 9, and 8 nm, respectively. The calculated values are 8–13 meV higher than the experimental data, presumably because of neglecting Coulomb interaction between electrons and holes in the calculation, which reduces ΔE. Nevertheless, this simple calculation strongly supports that blue-shift in the PL spectra of the ultrafine ZnO nanorods results from the diameter-dependent quantum-confinement effect.

PL spectra of ZnO/Zn$_{0.8}$Mg$_{0.2}$O radial nanorod heterostructures with various Zn$_{0.8}$Mg$_{0.2}$O overlayer thicknesses were measured in order to investigate the effect of the Zn$_{0.8}$Mg$_{0.2}$O layer thickness on the PL properties. Compared with the PL spectrum of 9-nm thick ZnO nanorods, as shown in Figure 5.12(b), PL spectra of

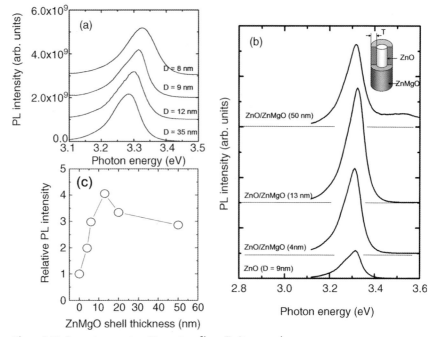

Figure 5.12 Room-temperature PL spectra of bare ZnO nanorods and ZnO/Zn$_{0.8}$Mg$_{0.2}$O radial nanorod heterostructures. (a) PL spectra of ZnO nanorods with a different mean diameter of 8, 9, 12, and 35 nm. (b) PL spectra of ZnO/Zn$_{0.8}$Mg$_{0.2}$O radial nanorod heterostructures depending on the Zn$_{0.8}$Mg$_{0.2}$O shell-layer thickness. (c) Normalized PL intensity of near-band-edge emission depending on the Zn$_{0.8}$Mg$_{0.2}$O shell-layer thickness.

ZnO/Zn$_{0.8}$Mg$_{0.2}$O radial nanorod heterostructures did not show any significant changes in the overall shape and dominant PL peak position with increasing the Zn$_{0.8}$Mg$_{0.2}$O layer thickness. Only for the radial nanorod heterostructures with a 50-nm thick Zn$_{0.8}$Mg$_{0.2}$O layer, a new emission peak was observed at 3.53 eV, corresponding to the near-band-edge emission of the Zn$_{0.8}$Mg$_{0.2}$O layer [43]. This result implies that the Zn$_{0.8}$Mg$_{0.2}$O layer coating on ZnO nanorods did not induce the formation of an intermediate alloy layer between ZnO and Zn$_{0.8}$Mg$_{0.2}$O, as similar behavior was observed for thin-film heterostructures [45].

The deposition of a Zn$_{0.8}$Mg$_{0.2}$O layer significantly enhanced the integrated PL intensity without changing the PL peak position. As shown in Figure 5.12(c), the PL intensity of the radial nanorod heterostructures increases with Zn$_{0.8}$Mg$_{0.2}$O layer thickness up to 13 nm, for a 13-nm thick Zn$_{0.8}$Mg$_{0.2}$O layer, exhibiting four times stronger than that of bare ZnO nanorods. This behavior can be explained in terms of reduction of surface states that may be sources of nonradiative recombination or radiative deep-level transitions [51, 52]. Bare ZnO nanorods have a high surface/volume ratio, and the surface effect can be dominant for nanorods with a small diameter. For a radial nanorod heterostructure with a clean interface, however, a $Zn_{0.8}Mg_{0.2}O$ shell layer confines carriers in the ZnO core material due to its larger bandgap energy, hence suppressing the surface state effect. In addition, the shell layer offers an increase in excitation volume. Accordingly the integrated PL intensity increases with increasing the shell-layer thickness up to 13 nm. However, further increase of shell-layer thickness leads to a decrease in the PL intensity by generating structural defects and another radiative transition at 3.53 eV in $Zn_{0.8}Mg_{0.2}O$.

$ZnO/Zn_{0.8}Mg_{0.2}O$ radial nanorod heterostructures exhibited further interesting optical properties due to enhanced quantum efficiency. Figure 5.13(a) presents typical PL spectra of ZnO (diameter of 9 nm)/$Zn_{0.8}Mg_{0.2}O$ (thickness of 13 nm) radial nanorod heterostructures measured at various temperatures of 10–293 K. In general, PL peak intensity at room temperature is one or two orders of magnitude lower than that at 10 K. However, the room-temperature PL intensity for the radial nanorod heterostructures was as high as 30% of the PL intensity at 10 K.

Figure 5.13(b) shows the normalized integrated PL intensities for thick (diameter of 35 nm) and ultrafine (diameter of 35 nm) ZnO nanorods and the $ZnO/Zn_{0.8}Mg_{0.2}O$ radial nanorod heterostructures. Thick ZnO nanorods exhibited a huge decrease in PL intensity with increasing temperature; the PL intensity ratio ($I_{R.T.}/I_{10\,K}$) was 0.015. For ultrafine ZnO nanorods, thermal quenching in luminescence was suppressed, so that a higher PL intensity ratio of 0.05 was observed. This results from enhanced electron and hole wavefunction overlap due to carrier localization [53]. As mentioned above, better optical qualities with $I_{R.T.}/I_{10\,K} = 0.3$ were achieved by capping a $Zn_{0.8}Mg_{0.2}O$ layer, presumably due to further reduced thermal quenching by passivation of surface nonradiative recombination centers.

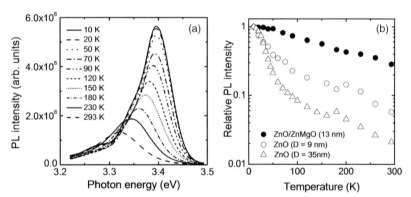

Figure 5.13 (a) Temperature-dependent PL spectra of ZnO (diameter 9 nm)/$Zn_{0.8}Mg_{0.2}O$ (thickness 13 nm) nanorod heterostructures measured in the temperature range from 10 to 293 K. (b) Spectrally integrated PL intensity normalized with the PL intensity at 10 K as a function of temperature for thick (diameter 35 nm) and ultrafine (diameter 9 nm) ZnO nanorods and the $ZnO/Zn_{0.8}Mg_{0.2}O$ radial nanorod heterostructures.

Meanwhile, if the shell-layer thickness for the nanorod heterostructures is controlled very accurately, multishell nanorod heterostructures exhibiting quantum-confinement effect can be fabricated. To observe quantum confinement of charge carriers in radial nanorod heterostructures for wide-bandgap materials such as ZnO and GaN, the QW shell-layer thickness must be smaller than a few nm [42, 54]. Moreover, the lattice mismatches between core and shell-layer components for the radial nanorod heterostructures are so small that interfacial strain should not be induced because the strain subsequently leads to formation of numerous crystal defects including misfit dislocations at the interface [37]. The exciting challenges to overcome such problems inspire us to explore quantum confinement in novel radial nanorod heterostructures. Here, the fabrication of ZnO/$Zn_{0.8}Mg_{0.2}$O core/multishell nanorod quantum structures and observation of quantum-confinement effect in the multishell quantum structures are described [6, 55].

$Zn_{0.8}Mg_{0.2}$O/ZnO/$Zn_{0.8}Mg_{0.2}$O radial nanorod SQWs were prepared on ZnO core nanorods using catalyst-free MOVPE as described in Figure 5.14. In this structure, carriers can be confined in a SQW, resulting in a blue-shift of a dominant excitonic emission in PL spectra due to the quantum-confinement effect. Such distinctive radial nanorod quantum structures could be realized by precise control of ZnO QW and $Zn_{0.8}Mg_{0.2}$O QB layer thicknesses in subnanometer scale. Almost defect-free interfaces in both axial and radial nanorod heterostructures have been observed using high-resolution transmission electron microscopy.

Radial nanorod SQWs with different ZnO QW shell thicknesses (L_W) were prepared in order to investigate a blue-shift of a dominant PL emission depending on QW layer thickness (QW width in the energy diagram shown in Figure 5.14). Bare ZnO nanorods exhibited a mean diameter of 26 nm and a length of 2.5 μm. The

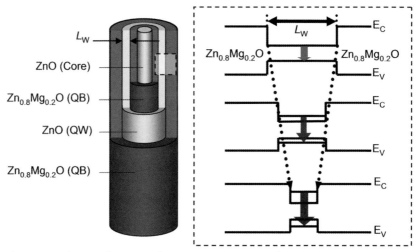

Figure 5.14 Schematic illustration of a ZnO/$Zn_{0.8}Mg_{0.2}$O coaxial nanorod SQW and its band diagrams for different ZnO QW shell widths (L_W).

continuous growths of the ZnO QW and $Zn_{0.8}Mg_{0.2}O$ QB shell layers lead to an increase in the radial nanorod diameter to 51, 52, and 58 nm for $Zn_{0.8}Mg_{0.2}O$ QB shell-layer growth times of 30, 60, 180 s, respectively. Averaging the diameters of the individual radial nanorod heterostructures from microscopic images indicates that the average growth rates of the ZnO QW and $Zn_{0.8}Mg_{0.2}O$ QB shell layers were roughly 0.25 and 0.33 Å/s, respectively.

PL spectra of the $ZnO/Zn_{0.8}Mg_{0.2}O$ core/shell radial nanorod heterostructures and core/multishell SQWs were measured at 10 K in order to compare the PL peaks from the heterostructures and SQWs. As shown in Figure 5.15(a), the PL spectra of the $ZnO/Zn_{0.8}Mg_{0.2}O$ core/shell radial nanorod heterostructures exhibited two dominant PL peaks at 3.364 and 3.353 eV corresponding to excitonic emissions from ZnO core nanorods ($I_2^{ZnO(core)}$) and $Zn_{0.8}Mg_{0.2}O$ (I^{ZnMgO}) shell layers, respectively. Meanwhile, for the $ZnO/Zn_{0.8}Mg_{0.2}O$ radial nanorod SQWs, a new PL peak was observed and the PL peak position blue-shifted with decreasing ZnO (QW) shell thickness. Figures 5.15(b)–(e) show the PL spectra of the nanorod SQWs with the ZnO QW shell-layer thickness varied from 8 to 45 Å and a $Zn_{0.8}Mg_{0.2}O$ QB shell-layer thickness fixed at 120 Å. When the ZnO (QW) shell thickness was 45 Å, a new PL peak marked by $I^{ZnO(QW)}$ was observed at 3.382 eV, indicating the blue-shift of 18 meV from the $I_2^{ZnO(core)}$ PL emission at 3.364 eV. As the ZnO QW width decreased from 45 to 8 Å, the $I^{ZnO(QW)}$ peak energy in Figures 5.15(b)–(e) blue-shifted up to 3.467 eV. The blue-shift was as large as 86 meV for the thinnest QW of 8 Å. Such a blue-shift of the PL emission peak can be well understood in terms of quantum size effect in the radial nanorod SQWs.

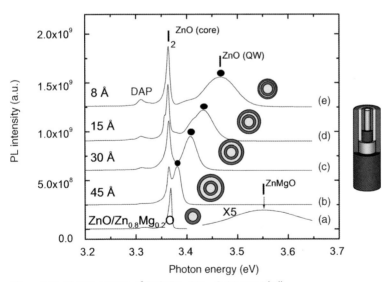

Figure 5.15 10-K PL spectra of $ZnO/Zn_{0.8}Mg_{0.2}O$ (a) core/shell nanorod heterostructures and (b)–(e) core/multishell nanorod quantum structures with different ZnO (QW) shell widths of 45, 30, 15, 8 Å.

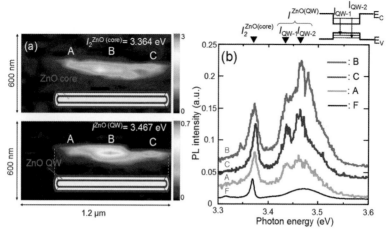

Figure 5.16 (a) Monochromatic PL images of an individual ZnO/Zn$_{0.8}$Mg$_{0.2}$O core/multishell nanorod SQW measured at different emission energies of 3.364 ($I_2^{ZnO(core)}$) and 3.467 ($I^{ZnO(QW)}$) eV. In the ZnO/Zn$_{0.8}$Mg$_{0.2}$O core/multishell nanorod SQWs, ZnO (QW) and Zn$_{0.8}$Mg$_{0.2}$O shell widths are 8 and 120 Å, respectively. (b) Near-field PL spectra taken at different positions marked by A, B, and C in (a). The black solid curve is the far-field spectrum.

The quantum effect creates quantized sublevel states in the QWs and increases quantized energy levels with decreasing the QW width.

Furthermore, spatially resolved PL images and spectra of a single ZnO/Zn$_{0.8}$Mg$_{0.2}$O radial nanorod SQWs were measured using near-field scanning optical microscopy (NSOM) at 10 K. Figure 5.16(a) and (b) show the near-field images and spectra of the radial nanorod SQWs with a ZnO well-layer thickness (QW) of 8 Å and a Zn$_{0.8}$Mg$_{0.2}$O (QB) of 120 Å. As shown in Figure 5.16(a), both near-field PL images taken at different emission energies of 3.364 ($I_2^{ZnO(core)}$) and 3.467 eV ($I^{ZnO(QW)}$) are nearly the same, only with the exception of topographically obtained emission intensity. Furthermore, from the near-field PL spectra collected from the different positions as described by A, B and C in Figure 5.16(a), the PL peak energies of 3.364 eV ($I_2^{ZnO(core)}$) and 3.467 eV ($I^{ZnO(QW)}$) are consistent with those of far-field PL spectrum, as shown in Figure 5.16(b). These results strongly suggest good homogeneity of the ZnO QW shell layer along the entire axial direction of radial nanorod SQWs.

It is worthwhile noting that the $I^{ZnO(QW)}$ peak at 3.467 eV in the far-field PL spectrum is separated into two peaks of 3.435 (I_{QW-1}) and 3.460 (I_{QW-2}) eV in the near-field PL spectra as shown in Figure 5.16(b). The discrete energy levels for the core/multishell nanorod quantum structures were observed at an excitation power density of 11 W/cm^2. Similar behavior on the peak splitting was previously observed for axial nanorod QWs, and was attributed to the transitions from ground an excited sublevels in the SQW [7]. Accordingly, the I_{QW-1} and I_{QW-2} emission lines are associated with the recombination of the ground state and the first excited state as depicted in the right upper side of Figure 5.16(b).

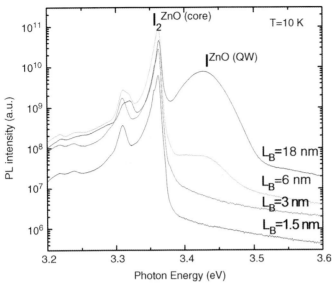

Figure 5.17 Low-temperature PL spectra of ZnO/Mg$_{0.2}$Zn$_{0.8}$O radial nanorod SQWs with various barrier layer thicknesses (L_B) of 1.5, 3, 6, and 18 nm.

PL spectra of the radial nanorod SQWs with various QB layer thicknesses were measured to investigate the effect of QB layer thickness on the quantum confinement of the carriers. Figure 5.17 shows 10-K PL spectra of the radial nanorod SQWs with a core ZnO diameter of 50 nm, a ZnO QW layer thickness of 1.5 nm, and various QB layer thicknesses of 1.5, 3, 6, and 18 nm. For a QB layer thickness of 18 nm, a strong PL peak was observed at 3.427 eV, resulting from the quantum-confinement effect in a single ZnO QW layer. The intensity of the PL peak decreased significantly with decreasing the QB layer thickness to 6 nm, and then totally disappeared for QB layer thicknesses of 1.5 and 3 nm. This suggests that a QB layer thickness of 6 nm is too thin to confine the carriers or that the QB layers are not uniformly covered with holes present resulting in carrier leakage.

Further optical properties of ZnO/Zn$_{0.8}$Mg$_{0.2}$O radial nanorod SQWs were investigated by measuring their temperature-dependent PL spectra in the range of 10 K–300 K. Figure 5.18 shows the temperature-dependent PL spectra of ZnO/Zn$_{0.8}$Mg$_{0.2}$O radial nanorod SQWs with an L_W of 15 Å. The energy positions of donor bound exciton peak $I_2^{ZnO(core)}$ and $I^{ZnO(QW)}$ decreased with bandgap energy shrinkage, and the PL peak intensities decreased with increasing temperature. However, thermal quenching behavior of $I^{ZnO(QW)}$ was different from that of $I_2^{ZnO(core)}$: the $I_2^{ZnO(core)}$ peak intensity decreased drastically and disappeared at temperatures above 120 K, whereas the $I^{ZnO(QW)}$ peak quenched slowly and survived at 300 K. The low thermal quenching of the $I^{ZnO(QW)}$ peak presumably results from the quantum-confinement effect of carriers in the QW, and also indicating an excellent optical characteristic of radial nanorod quantum

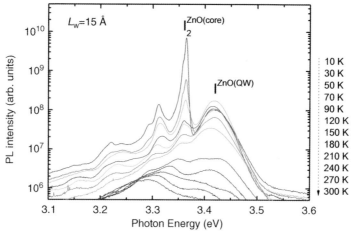

Figure 5.18 Temperature-dependent PL spectra of ZnO/Zn$_{0.8}$Mg$_{0.2}$O radial nanorod SQWs with L_W of 15 Å.

structures due to clean ZnO/Zn$_{0.8}$Mg$_{0.2}$O interfaces because PL quenching is also related to interfacial defects.

5.4
Conclusions

This chapter has reviewed the controlled heteroepitaxial growth and structural and optical characteristics of ZnO-based nanorod heterostructures and quantum structures for nanophotonic device applications. By using catalyst-free MOVPE, the thicknesses of QW and QB layers in ZnO/Zn$_{1-x}$Mg$_x$O nanorod heterostructures were controlled at an atomic-layer scale, while also achieving accurate composition control of the Zn$_{1-x}$Mg$_x$O alloy layers. The simple but precise thickness control allows heteroepitaxial growth of well-defined potential profiles resulting in nanosized QW structures in individual nanorods that are tunable through quantum confinement. The optical properties of the ZnO/Zn$_{1-x}$Mg$_x$O nanorod quantum structures are also described. In particular, both axial and radial nanorod quantum structures exhibited systematic PL peak shifts due to the quantum-confinement effect. More generally, the simple 'bottom up' epitaxial approach might readily be expanded to create many other heteroepitaxial semiconductor nanorods, offering important opportunities for the fabrication of nanoscale photonic devices.

Although this chapter has focused on ZnO-based nanomaterials and nanostructures for nanophotonic device applications, several reports have been published on ZnO nanorod-based optoelectronic and photonic device applications. Electroluminescent devices based on semiconductor nanorods and nanowires have recently been reported [56]. N-type ZnO nanorods and ZnO/GaN nanorod heterostructures, vertically aligned on p-GaN substrates, were directly configured to vertical EL

Figure 5.19 Position-controlled ZnO nanorod (left) and nanotube (right) arrays grown by catalyst-free MOVPE. Both nanostructures were prepared on Si substrates. The ZnO nanorod array (left) was selectively grown only on tips of GaN micropyramids and the nanotube array (right) grown along the circumferences of SiO_2 hole patterns.

devices [9, 57]. In particular, vertical light-emitting diode arrays based on the radial nanorod quantum structures are quite interesting because the vertical nanorod LEDs could offer essential building blocks for many other nanophotonic devices. Despite the recent progress in the preparation of nanorod heterostructures and their nanophotonic device applications, challenges still remain. The first challenge faced by the current ZnO nanorod synthesis methods is controlling the positions of the nanorods. Although self-assembly into complex structures and device architectures [58] or selective growth for position control of vertically aligned ZnO nanorod arrays [39, 40] have recently been reported (Figure 5.19) [59, 60], more reliable and reproducible position controls are required for practical applications of many nanophotonic devices. The second challenge for ZnO-based nanostructures is growing p-type ZnO nanorods and making p-n junction devices. Recent reports on p-type ZnO nanorods [24, 61] and p–n junction devices indicate that the quality of the nanorods is not yet sufficient for fabricating high-performance ZnO nanorod-based LEDs. If these problems can be resolved, ZnO-based nanorod heterostructures would present many fabrication possibilities for sophisticated nanophotonic devices as well as for simple LEDs.

Acknowledgements

This work was financially supported by the National Creative Research Initiative Project (R16-2004-004-01001-0) of the Korea Science and Engineering Foundations (KOSEF). The author would like to thank group members in the National Creative Research Initiative Center for Semiconductor Nanorods and Prof. M. Kim (Seoul National University) for TEM measurements and fruitful discussion.

References

1 Pauzauskie, P.J. and Yang, P. (2006) *Materials Today*, **9**, 36.
2 Duan, X.F., Huang, Y., Cui, Y., Wang, J.F. and Lieber, C.M. (2001) *Nature*, **409**, 66.
3 Yatsui, T., Sangu, S., Kawazoe, T., Ohtsu, M., An, S.J., Yoo, J. and Yi, G.-C. (2007) *Applied Physics Letters*, **90**, 223110.
4 Park, W.I., Yi, G.-C., Kim, M. and Pennycook, S.J. (2003) *Advanced Materials*, **15**, 526.
5 Park, W.I., An, S.J., Yang, J.L., Yi, G.-C., Hong, S., Joo, T. and Kim, M. (2004) *The Journal of Physical Chemistry. B*, **108**, 15457.
6 Jang, E.S., Bae, J.Y., Yoo, J., Park, W.I., Kim, D.W., Yi, G.-C., Yatsui, T. and Ohtsu, M. (2006) *Applied Physics Letters*, **88**, 023102.
7 Yatsui, T., Ohtsu, M., Yoo, J., An, S.J. and Yi, G.-C. (2005) *Applied Physics Letters*, **87**, 033101.
8 Qian, F., Gradečak, S., Li, Y., Wen, C.Y. and Lieber, C.M. (2005) *Nano Letters*, **5**, 2287.
9 Park, W.I. and Yi, G.-C. (2004) *Advanced Materials*, **16**, 87.
10 Kim, H.-M., Cho, Y.-H., Lee, H., Kim, S.I., Ryu, S.R., Kim, D.Y., Kang, T.W. and Chung, K.S. (2004) *Nano Letters*, **4**, 1059.
11 Kikuchi, A., Kawai, M., Tada, M. and Kishino, K. (2004) *Japanese Journal of Applied Physics*, **43**, L1524.
12 Sun, Y., Cho, Y.H., Kim, H.M. and Kang, T.W. (2005) *Applied Physics Letters*, **87**, 1.
13 Jeong, M.C., Oh, B.Y., Ham, M.H., Lee, S.W. and Myoung, J.M. (2007) *Small*, **3**, 568.
14 Björk, M.T., Ohlsson, B.J., Sass, T., Persson, A.I., Thelander, C., Magnusson, M.H., Deppert, K., Wallenberg, L.R. and Samuelson, L. (2002) *Applied Physics Letters*, **80**, 1058.
15 Thelander, C., Mårtensson, T., Björk, M.T., Ohlsson, B.J., Larsson, M.W., Wallenberg, L.R. and Samuelson, L. (2003) *Applied Physics Letters*, **83**, 2052.
16 Hu, Y., Xiang, J., Liang, G., Yan, H. and Lieber, C.M. (2008) *Nano Letters*, **8**, 925.
17 Huang, Y., Duan, X.F. and Lieber, C.M. (2005) *Small*, **1**, 142.
18 Sirbuly, D.J., Law, M., Pauzauskie, P., Yan, H., Maslov, A.V., Knutsen, K., Ning, C.Z., Saykally, R.J. and Yang, P. (2005) *Proceedings of the National Academy of Sciences of the United States of America*, **102**, 7800.
19 An, S.J., Chae, J.H. and Yi, G.-C. (2008) *Applied Physics Letters*, **92**, 121108.
20 Yi, G.-C., Wang, C. and Park, W.I. (2005) *Semiconductor Science and Technology*, **20**, S22.
21 Ohtsu, M., Kobayashi, K., Kawazoe, T., Sangu, S. and Yatsui, T. (2002) *IEEE Journal of Selected Topics in Quantum Electronics*, **8**, 839.
22 Park, W.I., Kim, D.-W., Jung, S.W. and Yi, G.-C. (2006) *International Journal of Nanotechnology*, **3**, 372.
23 Yoo, J. and Yi, G.-C. (2006) in *Undoped and Doped ZnO Nanorods*, Pennsylvania State University, Pennsylvania, (TMS).
24 Xiang, B., Wang, P.W., Zhang, X.Z., Dayeh, S.A., Aplin, D.P.R., Soci, C., Yu, D.P. and Wang, D.L. (2007) *Nano Letters*, **7**, 323.
25 Yuan, G.-D. and Lee, S.-T. (2008) *Advanced Materials*, **20**, 168.
26 Lee, C.J., Lee, T.J., Lyu, S.C., Zhang, Y., Ruh, H. and Lee, H.J. (2002) *Applied Physics Letters*, **81**, 3648.
27 Yonenaga, I. (2001) *Physica B*, **308–310**, 1150.
28 Gudiksen, M.S., Lauhon, L.J., Wang, J., Smith, D.C. and Lieber, C.M. (2002) *Nature*, **415**, 617.
29 Solanki, R., Huo, J., Freeouf, J.L. and Miner, B. (2002) *Applied Physics Letters*, **81**, 3864.
30 Park, W.I., Kim, D.H., Jung, S.W. and Yi, G.C. (2002) *Applied Physics Letters*, **80**, 4232.
31 Kim, H.M., Kang, T.W. and Chung, K.S. (2003) *Advanced Materials*, **15**, 567.
32 Noborisaka, J., Motohisa, J. and Fukui, T. (2005) *Applied Physics Letters*, **86**, 213102.

33 Wang, X., Sun, X., Fairchild, M. and Hersee, S.D. (2006) *Applied Physics Letters*, **89**, 233115.
34 Lee, W., Jeong, M.C. and Myoung, J.M. (2004) *Acta Materialia*, **52**, 3949.
35 Park, W.I., Yi, G.C., Kim, M.Y. and Pennycook, S.J. (2002) *Advanced Materials*, **14**, 1841.
36 Park, W.I., Jun, Y.H., Jung, S.W. and Yi, G.C. (2003) *Applied Physics Letters*, **82**, 964.
37 An, S.J., Park, W.I., Yi, G.C., Kim, Y.J., Kang, H.B. and Kim, M. (2004) *Applied Physics Letters*, **84**, 3612.
38 Park, C.J., Choi, D.-K., Yoo, J., Yi, G.-C. and Lee, C.J. (2007) *Applied Physics Letters*, **90**, 083107.
39 Hong, Y.J., An, S.J., Jung, H.S., Lee, C.-H. and Yi, G.-C. (2007) *Advanced Materials*, **19**, 4416.
40 Hong, Y.J., Jung, H.S., Yoo, J., Kim, Y.-J., Lee, C.-H., Yi, G.-C. and Kim, M. *Advanced Materials*, (in press).
41 Park, W.I., Jung, S.W., Jun, Y. H. and Yi, G.-C., in *Heteroepitaxial ZnO/ZnMgO quantum structures in nanorods*, Edinburgh, Scotland, UK, 2002 (Institute of Physics).
42 Mohan, P., Motohisa, J. and Fukui, T. (2006) *Applied Physics Letters*, **88**, 133105.
43 Park, W.I., Yi, G.-C. and Jang, H.M. (2001) *Applied Physics Letters*, **79**, 2022.
44 Ohtomo, A., Kawasaki, M., Ohkubo, I., Koinuma, H., Yasuda, T. and Segawa, Y. (1999) *Applied Physics Letters*, **75**, 980.
45 Ohtomo, A., Shiroki, R., Ohkubo, I., Koinuma, H. and Kawasaki, M. (1999) *Applied Physics Letters*, **75**, 4088.
46 Jung, S.W., Park, W.I., Cheong, H.D., Yi, G.C., Jang, H.M., Hong, S. and Joo, T. (2002) *Applied Physics Letters*, **80**, 1924.
47 Meyer, B.K., Alves, H., Hofmann, D.M., Kriegseis, W., Forster, D., Bertram, F., Christen, J., Hoffmann, A., Straßburg, M., Dworzak, M., Haboeck, U. and Rodina, A.V. (2004) *Physica Status Solidi B-Basic Research*, **241**, 231.
48 Kim, C., Park, W.I. and Yi, G.-C. (2006) *Applied Physics Letters*, **89**, 113106.
49 Teichert, C., Lagally, M.G., Peticolas, L.J., Bean, J.C. and Tersoff, J. (1996) *Physical Review B-Condensed Matter*, **53**, 16334.
50 Zervos, M. and Feiner, L.F. (2004) *Journal of Applied Physics*, **95**, 281.
51 Wong, E.M. and Searson, P.C. (1999) *Applied Physics Letters*, **74**, 2939.
52 Zhou, H., Alves, H., Hofmann, D.M., Kriegseis, W., Meyer, B.K., Kaczmarczyk, G. and Hoffmann, A. (2002) *Applied Physics Letters*, **80**, 210.
53 Cho, Y.H., Kwon, B.J., Barjon, J., Brault, J., Daudin, B., Mariette, H. and Dang, L.S. (2002) *Applied Physics Letters*, **81**, 4934.
54 Lin, H.M., Chen, Y.L., Yang, J., Liu, Y.C., Yin, K.M., Kai, J.J., Chen, F.R., Chen, L.C., Chen, Y.F. and Chen, C.C. (2003) *Nano Letters*, **3**, 537.
55 Bae, J.Y., Yoo, J. and Yi, G.-C. (2006) *Applied Physics Letters*, **89**, 173114.
56 Svensson, C.P.T., Mårtensson, T., Trägårdh, J., Larsson, C., Rask, M., Hessman, D., Samuelson, L. and Ohlsson, J. (2008) *Nanotechnology*, **19**, 305201.
57 An, S.J. and Yi, G.-C. (2007) *Applied Physics Letters*, **91**, 123109.
58 Thelander, C., Agarwal, P., Brongersma, S., Eymery, J., Feiner, L.F., Forchel, A., Scheffler, M., Riess, W., Ohlsson, B.J., Gösele, U. and Samuelson, L. (2006) *Materials Today*, **9**, 28.
59 Hong, Y.J., An, S.J., Jung, H.S., Lee, C.-H and Yi, G.-C. (2007) *Advanced Materials*, **19**, 4416.
60 Hong, Y.J., Jung, H.S., Yoo, J., Kim, Y.-J., Lee, C.-H. and Yi, G.-C. (2009) *Advanced Materials*, **21**, 222.
61 Yuan, G.D., Zhang, W.J., Jie, J.S., Fan, X., Zapien, J.A., Leung, Y.H., Luo, L.B., Wang, P.F., Lee, C.S. and Lee, S.T. (2008) *Nano Letters*, **8**, 2591.

6
Lithography by Nanophotonics

Ryo Kuroda, Yasuhisa Inao, Shinji Nakazato, Toshiki Ito, Takako Yamaguchi, Tomohiro Yamada, Akira Terao, and Natsuhiko Mizutani

6.1
Introduction

Image information becomes more high-definition and larger capacity in imaging instruments such as digital cameras and digital videos. Information-recording medium is hoped to be miniaturized and made lighter in the instruments. In mobile personal computers, the recording medium is being changed from hard disk drives (HDD) to solid-state drives (SSD) of the semiconductor memory base. The storage capacity of the current semiconductor memory such as a flash memory is not enough, and further increases in capacity are being demanded.

Therefore, a further resolution improvement is needed in the semiconductor exposure device as a manufacturing tool. In photolithography, microprocessing resolution R is expressed as Rayleigh's expression;

$$R = (k_1 \cdot \lambda)/\mathrm{NA}$$

where k_1 is called a process coefficient, and it is decided by the characteristic of the photoresist material. It is said that the lower limit in k_1 has a value of 0.3. λ is the wavelength of light used for the exposure. The numerical aperture NA is twice the focal length divided by the diameter of the lens. The microprocessing resolution R is decided by the wavelength of light λ and the numerical aperture NA of the lens that projects mask patterns to the substrate. From this expression, it is understood that the wavelength of light λ should be shorter and the numerical aperture NA of the lens should be higher to improve the microprocessing resolution.

The immersion ArF lithography system with exposure light of short wavelength ($\lambda = 193$ nm), high NA (>1) optics and double-patterning process have been developed. Another lithography system that uses EUV light ($\lambda = 13.5$ nm) as a light source is also being developed. In these lithography systems, the light source, the exposure optics, the machine driving device and the measurement control device are all extremely sophisticated, and the cost of the whole system becomes very high.

Nanophotonics and Nanofabrication. Edited by Motoichi Ohtsu
Copyright © 2009 WILEY-VCH Verlag GmbH & Co. KGaA, Weinheim
ISBN: 978-3-527-32121-6

On the other hand, various nanostructure devices are proposed in the progress of nanotechnology. In general, nanodevices have a tendency of being small but producing numerous products. Since the high-cost manufacturing tool as mentioned above cannot be used in putting such nanodevices to practical use, a low-cost nanoprocessing tool is sought. One of the candidates of such a tool is the nanoimprint lithography system. A principle of nanoimprint lithography is that softened resin is put into a mold and stiffened. The effort toward shortening the resin filling time and decreasing defects by the resin flake is continued aiming at practical use.

Under these background conditions, research on the microprocess using an optical near-field localized around a nanoaperture of the probe tip was carried out [1–3]. Moreover, the mask type of optical near-field lithography using a photomask with plural nanoapertures was proposed to improve the processing throughput. Continuous research and development of a photomask, a photoresist and resist process that are appropriate for the optical near-field lithography have been performed for practical use [4–6]. Optical near-field lithography is improving the NA of the optics in the photolithography to the limit, as explained later, and a new photolithography technology that applies the nanophotonics to microfabrication.

6.2
Principle of the Optical Near-Field Lithography

Figure 6.1 shows the principle of the optical near-field lithography. A near-field photomask consists of predetermined patterns of apertures of a width of 10–100 nm that is smaller than the wavelength of the exposure light in opaque film made of metal, etc. The thickness of the film is about 50 nm. I-line light (wavelength 365 nm) of the mercury lamp as an exposure light source is irradiated from the back to the nanoapertures, and an optical near-field localized around the front side of the nanoapertures is generated (Figure 6.1(a)). The distance that the optical near-field

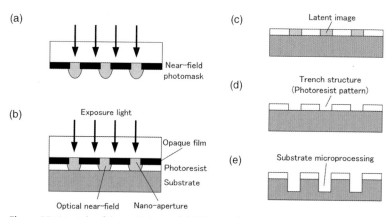

Figure 6.1 Principle of the optical near-field lithography.

spreads from the nanoapertures is almost equal to the size of the aperture width. Therefore, the near-field photomask is brought into contact with a photoresist on a wafer substrate within the distance (Figure 6.1(b)), and latent images of plural nanoapertures are simultaneously formed in the photoresist with the resolution of nanometer (Figure 6.1(c)). When the substrate is immersed in a developer, the exposed part of the (positive) photoresist dissolves and trench structures of the photoresist pattern are formed (Figure 6.1(d)). Based on these trench structures, the substrate is microprocessed in the following process (Figure 6.1(d)).

The extension size of this optical near-field is not limited by the wavelength of light (without diffraction limit), but depends on the size of the aperture. Therefore, the resolution of the optical near-field lithography depends on the size of the nanoaperture. The smaller the size of aperture becomes, the smaller pattern can be formed. This means that the shorter wavelength of light is not necessary for optical near-field lithography. Moreover, the aberration that becomes a problem of blurring the transferred pattern in conventional photolithography does not affect the optical near-field lithography with no lens but apertures. These features without an expensive light source and optics essentially make the optical near-field lithography a low-cost one.

Figure 6.2 shows a comparison of the optical system of the optical near-field photolithography with that of the conventional lithography. The image of reticle patterns was transferred onto a wafer with an imaging lens in the conventional photolithography as shown in Figure 6.2(a). The angle of the diffraction from the reticle pattern becomes large as the pattern size becomes small, and the imaging lens cannot collect the diffracted light. To solve this problem, off-axis illumination method was developed. The exposure light irradiates the reticle diagonally. The 0th-order transmitted light and the 1st-order diffraction light from the reticle interferes with each other, and this interference forms a latent image in the photoresist. If the reticle

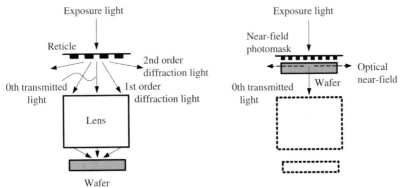

Figure 6.2 Comparison of the optical system of the optical near-field photolithography with that of the conventional lithography. (a) Conventional photolithography. (b) Optical near-field lithography.

pattern becomes small, however, the diffraction angle becomes larger and the diffraction light cannot enter the imaging lens and latent image cannot be formed. In order to fabricate finer patterns, a shorter wavelength of light and higher NA optics are required.

On the other hand, in optical near-field lithography, the size of the nanoaperture pattern of photomask is smaller than the wavelength of light as shown in Figure 6.2(b). The diffraction light does not radiate from the surface of the photomask, but exists only as an optical near-field localized around the surface. Then, all of the diffraction light of higher order including the localized near-field light can be used to form the latent image by bringing the surface of the photoresist into contact with the photomask. The optics of the optical near-field lithography can use all the information of the patterns in the photomask though it does not use the lens. In this sense, it can be called an optical system with ultimate ultrahigh NA.

Figure 6.3 shows the distribution of the electric field near the nanoslit apertures of an optical near-field photomask analyzed by a finite-difference time-domain (FDTD) method. The optical near-field generated on the output side of the nanoslit aperture is strong in the edge part of the aperture. The optical near-field extends circularly from this region sharply and the intensity of it decreases sharply. Figure 6.4 shows the dependence of the depth of the optical near-field on the aperture pitch. The maximum value of the intensity of the optical near-field at a certain depth is defined as I_{max} and the minimum value is defined as I_{min}. The contrast C in the depth is defined as $C = (I_{max} - I_{min})/(I_{max} + I_{min})$. The aperture ratio R is defined as $R =$ aperture width/aperture pitch. A contrast value of 0.5 must be maintained to form the latent pattern in the photoresist. The depth of the optical

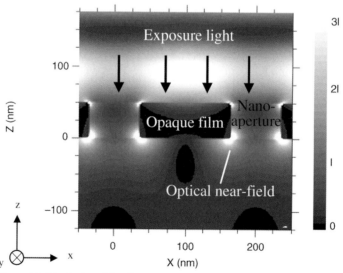

Figure 6.3 Distribution of the electromagnetic field near nanoslit apertures of an optical near-field photomask. The intensity of the incident light is expressed as I.

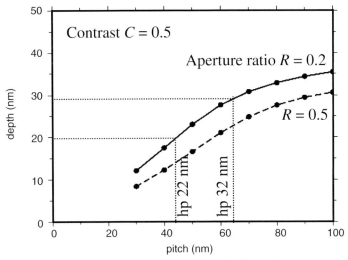

Figure 6.4 Dependence of depth of the optical near-field on pitch of the patterns. Contrast value of 0.5 must be kept necessary to form the latent pattern in the photoresist. The depth becomes larger in inverse proportion to aperture ratio.

near-field that can be used for the exposure changes by the aperture ratio, and the appropriate depth is roughly 1/2 of the aperture pitch. Preferably, it is about 1/4 of the aperture pitch. Therefore, it is also preferable that the thickness of the thin photoresist is about 1/4 of the aperture pitch. For instance, about 10 nm or less of the photoresist thickness is suitable for fabrication of the pattern of 44 nm pitch.

Detailed analysis of the electromagnetic field around the aperture is explained as follows. If the width of the nanoslit aperture becomes smaller than 100 nm, namely about 1/3 or less of the wavelength of the light, the parallel polarization component of the electric field, i.e. TE-polarized light (y-direction component in Figure 6.5), hardly penetrates through the aperture because of the cutoff characteristics of a narrow waveguide. On the other hand, the orthogonal polarization component, i.e. TM-polarized light (the x-direction component in Figure 6.5), can penetrate, because the polarization component in the orthogonal direction excites surface plasmon polaritons in the slit. The surface waves generated on two boundary sides of the slit couple each other and propagate inside of such a narrow slit. Now, there are various directions of actual patterns in the near-field photomask, and it seems that the polarization of the electric field must be controlled corresponding to the directions of the patterns. However, it is not necessary to control the polarization component of the irradiating light because only an orthogonal polarization component penetrates selectively for the above reason.

If the aperture pitch becomes small, it seems that the optical near-fields generated from adjacent apertures overlap mutually, and the resolution decreases. Phases of the optical near-fields generated from adjacent apertures are opposite right under the opaque film between the adjacent apertures, as shown in Figure 6.5,

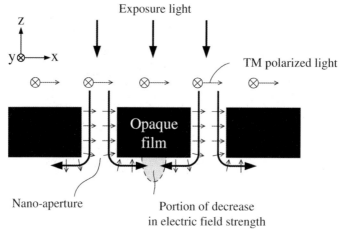

Figure 6.5 Overlap of the optical near-fields generated from adjacent apertures. The phases of the optical near-fields generated from adjacent apertures are opposite right under the opaque film between the adjacent apertures, and the electric field strength decreases.

and the optical near-field overlaps oppositely and the electric field strength decreases in that portion. Thus, the optical near-fields generated from the adjacent apertures function mutually to suppress the extension, and the contrast of the latent image increases. This function looks like the phase-shift mask technology in one of the resolution-enhancement technologies used in conventional photolithography. The phase-shift mask technology improves the contrast of the light intensity on the substrate by shifting the phase of the light that passes the adjacent apertures on the reticle by π.

6.3
Optical Near-Field Lithography System

The whole surface of the near-field photomask should be brought into contact with the photoresist within the localizing distance of the optical near-field. The near-field photomask is formed with a membrane that can deform elastically. A differential pressure of about 10 kPa is applied between the front side and the back side of the photomask. The membrane photomask deforms elastically and is brought into contact with the photoresist within the distance of 20 nm or less by a uniform force. This contact mechanism is so soft, unlike the contact exposure system in the conventional photolithography, that neither a deformation of the photoresist nor adhesion to that occurs. Mechanical control of the distance and the tilt of the two axes between the photomask and the photoresist is not needed because the membrane photomask deforms elastically and is brought into contact with the photoresist.

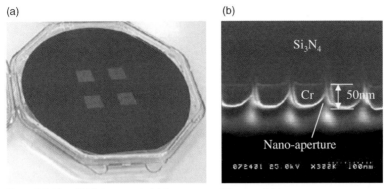

Figure 6.6 Near-field photomask. (a) Appearance of the near-field photomask. (b) SEM photograph in the section of the photomask.

The system cost of the optical near-field lithography using the membrane photomask can be reduced due to this.

Figure 6.6 shows the appearance of the near-field photomask and the SEM photograph in the section. The fabrication process of the near-field photomask is shown in Figure 6.7. Si_3N_4 film (500 nm in thickness) is formed on a Si substrate and opaque film (50-nm thick Cr film) is formed on the Si_3N_4 film. Predetermined nanoaperture patterns are fabricated in the opaque film by electron beam (EB) lithography or focused ion beam (FIB) fabrication. Finally, a part of the Si substrate that supports the thin Si_3N_4 film is removed by using anisotropic etching with a KOH solution, and a membrane structure of the photomask is formed.

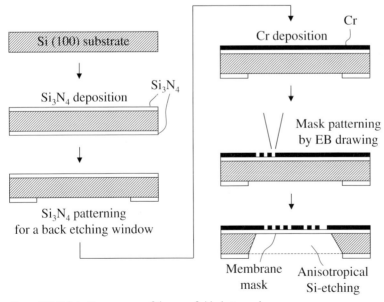

Figure 6.7 Fabrication process of the near-field photomask.

Figure 6.8 Appearance of a prototype device of optical near-field lithography.

As mentioned above, the photoresist should be thin according to the distance of localization of the optical near-field around the nanoaperture. On the other hand, the photoresist should have the dry etching resistance so that the latent image can be transferred to the lower layer substrate. To separate the function of the thin film and the dry etching resistance, we used a trilayer resist process; the upper layer is an ultrathin photoresist, the middle layer is thin spin on glass (SOG), and the lower layer is a resin. The patterns of shallow latent images in the upper layer are transferred to the thin SOG layer of excellent dry-etching resistance, and these are transferred in the lower layer further.

Figure 6.8 shows the appearance of a prototype device of optical near-field lithography [7]. The features of the prototype device are as follows. This device is compact, with a footprint of about $2\,m^2$. It has symmetric structure to compensate thermal expansion and temperature drift, and the hanging structure is introduced to avoid floor vibrations. In addition, it has a double clean structure, and the near-field photomask and the wafer are kept in a local clean environment by controlling a flow of clean air of the device inside.

Figure 6.9(a) shows the inside of the device. The i-line light (365 nm in wavelength) of a mercury lamp is introduced by the light guide as a light source for the exposure in the device, and irradiates the near-field photomask from the back as shown in Figure 6.9(b). Figure 6.9(c) is a close-up, observed from the side, of the part where the

Figure 6.9 (a) Inside the prototype device of the near-field lithography. (b) Illumination of i-line light for exposure from the back side of the near-field photomask. (c) Close-up from the side of the near-field photomask and the photoresist on the wafer. They are brought into contact with each other within the localizing distance of the optical near-field.

(a)

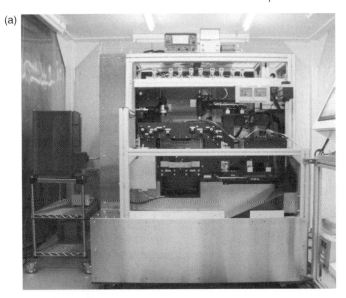

(b)

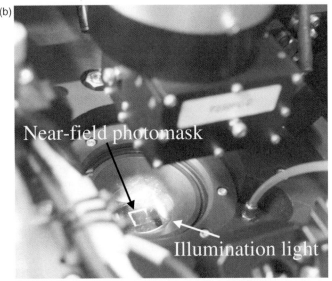

Near-field photomask

Illumination light

(c)

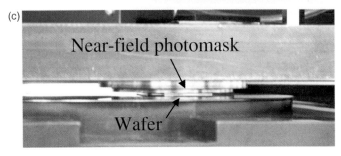

Near-field photomask

Wafer

near-field photomask and the photoresist on the wafer are brought into contact with each other within the localizing distance of the optical near-field.

The wafer to be exposed is held on an air-bearing stage, and a relative position of the photomask and the wafer is observed with a microscope. Using these mechanisms, the wafer is sequentially moved and exposed by a step-and-repeat method while observing a relative position for the photomask, and the patterns of the near-field photomask can be repeatedly transferred on the whole surface of the wafer.

6.4
Fabricated Patterns by Optical Near-Field Lithography

Several patterns fabricated by optical near-field lithography are shown in Figure 6.10.

Figure 6.10(a) shows a half-pitch 32-nm, high aspect photoresist pattern of 100 nm height. It is a pattern that the latent image of the photoresist in the upper layer is transferred into the lower resin by using the trilayer photoresist process.

Figure 6.10(b) shows the photoresist pattern of 50 nm width widely formed in the area of 5 mm × 5 mm on the substrate. However, there exist some defects in the pattern. These originate in defects of the near-field photomask in its manufacture. The manufacture of the photomask without defects is one of the problems that should be solved in the future.

Figure 6.10(c) shows the patterns formed on the whole surface of the 4-inch wafer with a step-and-repeat mechanism.

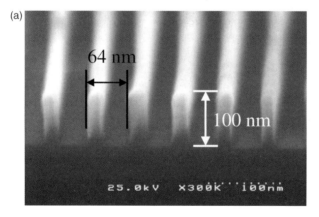

Figure 6.10 Several patterns fabricated by optical near-field lithography. (a) Half-pitch 32 nm, a high aspect pattern of 100 nm in height. (b) Patterns of 50 nm width widely formed in the area of 5 mm × 5 mm. (c) Patterns formed on the whole surface of the 4-inch wafer with a step-and-repeat mechanism. (d) Various two-dimensional patterns; a hole array, a dot array, a triangular lattice and a tetragonal lattice with periodic defects.

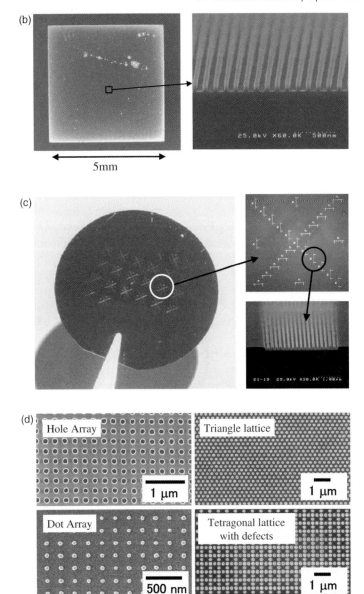

Figure 6.10 (Continued)

Figure 6.10(d) shows various two-dimensional patterns of a hole array, a dot array, a triangular lattice, and a tetragonal lattice with periodic defects that are related to a subwavelength structure optical element, a photonic crystal, and a nanostructure device of the quantum dot, etc.

6.5
Improvement of Resolution and Fabricated Ultrafine Patterns

It was shown that the structure of 100 nm size or less below the wavelength of the light for the exposure can be fabricated by optical near-field lithography as stated above. Our approaches to improve the resolution further are explained as follows.

First, a photoresist optimized for the optical near-field lithography has been developed. The chemically amplified photoresist is one of the highest-resolution photoresists. In the chemically amplified resist system, photogenerated acid diffuses and reacts with resist material in the process of postbaking, and a latent image is formed. Thus, the resolution is restricted by the diffusion length of the photogenerated acid. The diffusion length of the acid is estimated to be of the order of 10 nm. Therefore, resolution of the chemically amplified resist is not enough to fabricate a pattern of 10 nm order. To solve this problem, a photodeprotection resist system without diffusion has been recently developed [8]. In this system, the alkali-insoluble material changes to be alkali soluble after light exposure. The latent image does not become unclear because there is no diffusion. Figure 6.11 shows AFM images of hp 32-nm patterns that compare the photodeprotection resist system with the chemically amplified resist system.

Secondly, the photomask made of amorphous Si as the material of the opaque film was developed [9]. If the aperture pitch is reduced to fabricate a finer pattern, the oozing depth of the optical near-field decreases as shown in Figure 6.4, and it becomes difficult to form the finer pattern of the latent image in the photoresist. In the near-field photomask, a decrease in aperture ratio R can make the optical near-field deepen. Si can be processed into patterns of a finer and higher aspect compared with Cr, and the coefficient of Si at a wavelength of 365 nm is comparatively large, $\sim$2.6. Thus, Si has been chosen as the material of the opaque film to form as small an aperture width as possible. By optimizing the dry-etching conditions, apertures of 15 nm in width corresponding to about 1/3 of the aperture pitch of 44 nm were fabricated in the photomask, as shown in Figure 6.12.

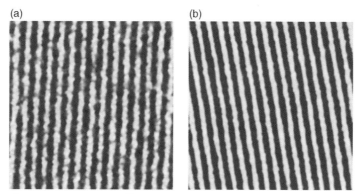

Figure 6.11 AFM images of hp 32-nm patterns using (a) the chemically amplified resist and (b) the photodeprotection resist.

6.5 Improvement of Resolution and Fabricated Ultrafine Patterns | 143

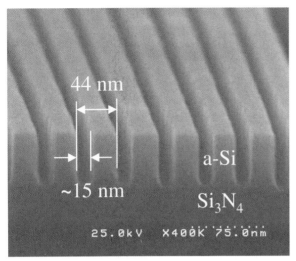

Figure 6.12 Near-field photomask for hp 22-nm patterning with apertures of 15 nm width corresponding to about 1/3 of the aperture pitch of 44 nm.

Lastly, the film thickness of the photoresist was optimized [10]. The optical near-field that spreads out from the aperture of the near-field photomask is converted into the propagation light, and it propagates in the photoresist. It is reflected on the surface of the substrate, and irradiates the front side of the photomask again. This reflected light interferes with the optical near-field around the nanoaperture of the photomask, and distribution of the optical near-field around the nanoaperture is modified. A kind of microresonator is formed between the photomask and the substrate. The graph in Figure 6.13 shows the contrast of the optical near-field versus total resist thickness. The contrast changes periodically as the phase relationship

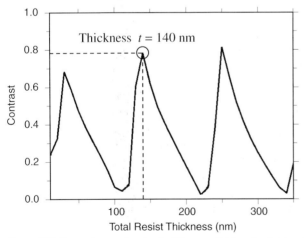

Figure 6.13 Dependence of contrast of the near-field on the total resist thickness.

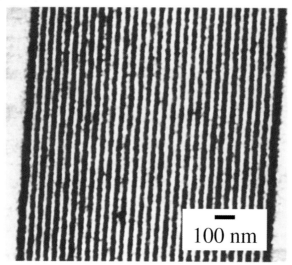

Figure 6.14 Fabricated pattern of half-pitch 22 nm.

between the near-field and the reflection light varies. Thus, the thickness of the lower resin, namely the distance between the photomask and the substrate was chosen as 140 nm so that the contrast of the optical near-field of the nanoaperture around the photomask might become the maximum.

The above three approaches have been made for higher resolution of the optical near-field lithography, and the fabricated pattern of half-pitch 22 nm is shown in Figure 6.14 [8]. The value of 22 nm is 1/16 or less of the wavelength (365 nm) of the light used for the exposure. This proved that optical near-field lithography has the high-resolution ability.

6.6
Summary

The prototype device of the optical near-field lithography has been developed, and hp 22-nm patterns could be successfully fabricated. This device is situated at the Tokyo University, and opened as a nanofabrication tool. Actually, a substrate of an X-ray diffraction grating has been fabricated by using this device. Low-cost and batch-processing nanofabrication technology, like optical near-field lithography, is appropriate to manufacture various nanodevices, as shown in Figure 6.15. Now, key approaches to improve further the performance of the optical near-field lithography technology will be explained as follows.

First, ultrathin-film photoresist of an atomic or molecular level will be necessary in future. The resolution of the optical near-field lithography depends on the optical near-field distribution around the nanoaperture of the photomask. Thinner photoresist improves the resolution because the contrast of the latent image becomes high

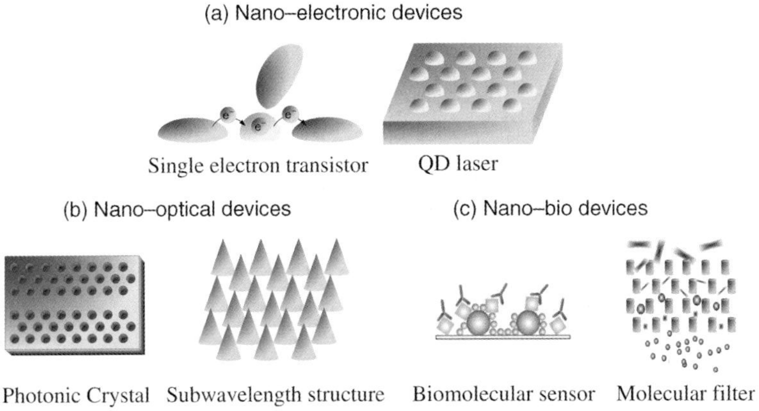

(a) Nano–electronic devices

Single electron transistor   QD laser

(b) Nano–optical devices   **(c) Nano–bio devices**

Photonic Crystal   Subwavelength structure   Biomolecular sensor   Molecular filter

Figure 6.15 Various nanodevices. (a) Nanoelectronic devices. (b) Nano-optical devices. (c) Nanobio devices.

as it approaches to the surface, whereas it becomes difficult to keep the etching resistance for the transfer to the lower layer when the photoresist becomes thin. Therefore, it is necessary to develop a new photoresist system that combines the surface exposure in the thickness at an atomic or molecular level with the transfer process after the exposure.

Secondly, it is necessary to elucidate the physics of the interface between the thin-film photomask and the photoresist. It is necessary to control the interface between the photomask and the photoresist at an atomic or molecular level so that the photomask can be repeatedly brought into contact with the photoresist within a distance of 10 nm or less and the photomask can be repeatedly peeled from the photoresist. Actually, it is confirmed that contact and peeling can be repeated more than 20 000 times after the surface of the photomask is treated by a fluorine-based material. However, it is not well understood what phenomenon occurs at the interface in contact and peeling. There is in the history of the photolithography technology the view that the contact method at the beginning of the photolithography moved to the projection method because of various problems of the photomask and the wafer substrate in the contact method. However, there is no method, except the proximity interaction that is used as a physical principle of the processing to improve the resolution of the ultrafine fabrication technology further. Optical near-field lithography is one of these methods that evolve the contact method in a new direction. This contact method should not be a conventional simple contact method, but a new contact method in which the contact boundary between the photomask and the substrate is controlled at an atomic or molecular level.

Lastly, it is necessary to investigate various pattern formations by multiple exposures and changing the exposure condition that make the most of the feature of the optical near-field lithography. The spatial distribution of the optical near-field can be changed even with the same photomask by changing the wavelength and the angle of incidence degree of the exposure light. In addition, it is possible to improve resolution and form complex patterns by multiple exposures.

In the optical near-field lithography technology explained above, the localization of the optical near-field was discussed. It is a quantitative revolution that ultrafine patterns are fabricated by using localized optical near-field. Recently, the phenomenon to expose photoresist by causing the reaction that does not take place because of a forbidden transition in usual light by using nonresonant process possible only by the optical near-field has been found in optical near-field lithography [11]. It is a qualitative revolution that the phenomenon that occurs only with an optical near-field. The research on nanophotonics will advance further, and new phenomena that relate to the optical near-field will be found in the future. Adding them, the optical near-field lithography will enter a new stage to qualitative revolutions.

Acknowledgment

A part of this research was jointly carried out with Prof. Otsu in Tokyo University in the entrusted business 'Practical use development of the nanomeasurement and the nanoprocessing technology' from the Ministry of Education, Culture, Sports, Science and Technology.

References

1 Betzig, E. and Trautman, K. (1992) *Science*, **257**, 189.
2 Nakajima, K., Mitsuoka, Y., Chiba, N., Muramatsu, H. and Ataka, T. (1995) *Proceedings of SPIE*, **2535**, 16.
3 Polonski, V., Yamamoto, Y., Kourogi, M., Fukuda, H. and Ohtsu, M. (1999) *Journal of Microscopy*, **194**, 545.
4 Japan Patent No. 03950532 (1997) and U. S. Patent No. 6171730.
5 Alkaisi, M.M., Blaikie, R.J., McNab, S.J., Cheung, R. and Cumming, D.R.S. (1999) *Applied Physics Letters*, **75**, 3560–3562.
6 Naya, M., Tsuruma, I., Tani, T. and Mukai, A. (2005) *Applied Physics Letters*, **86**, 201113.
7 Inao, Y., Nakazato, S., Kuroda, R. and Ohtsu, M. (2007) *Microelectronic Engineering*, **84**, 705–710.
8 Ito, T., Terao, A., Inao, Y., Yamaguchi, T. and Mizutani, N. (2007) *Proceedings of SPIE*, **6519**, 6519J.
9 Ito, T., Yamada, T., Inao, T., Yamaguchi, T., Mizutani, N. and Kuroda, R. (2006) *Applied Physics Letters*, **89**, 033113.
10 Yamaguchi, T., Yamada, T., Terao, A., Ito, T., Inao, Y., Mizutani, N. and Kuroda, R. (2007) *Microelectronic Engineering*, **84**, 690–693.
11 Kawazoe, T., Kobayashi, K., Takubo, S. and Ohtsu, M. (2005) *Journal of Chemical Physics*, **122**, 024715.

7
Nanopatterned Media for High-Density Storage
Hiroyuki Hieda

7.1
Introduction

Storage technology has underpinned development of the digital information society. Hard disk drives (HDD) have been used as main storage devices on personal computers, network servers, digital video recorders, and other consumer electronic devices. The data storage capacity of these devices has continued to increase. However, the storage technology is subject to a physical limit, namely, the thermal fluctuation limit.

In present-day magnetic storage, a magnetic mark is stored on a magnetic media as shown schematically in Figure 7.1. The recording layer consists of magnetic crystal particles and a nonmagnetic matrix surrounding the magnetic particles. This magnetic recording media is called 'granular structure media'. Because the nonmagnetic matrix divides the magnetic interaction between the magnetic crystal particles, the magnetic particles behave as single units of magnetic domains. The diameter of the magnetic crystal particles is critical for high-density recording. It is assumed that the blurred region between neighboring recorded magnetic marks is determined by the diameter of the particles. Therefore, to increase the density of recording marks while maintaining the signal-to-noise ratio (SNR), it has been necessary to reduce the size of magnetic particles. If the magnetic particle size is to be further reduced, we will encounter a physical limit, namely, the superparamagnetic limit. The thermal stability of a magnetic particle is determined by $K_u V$; K_u is the magnetic crystal anisotropy energy, V is the volume of the particle. It is thought that $K_u V$ should be larger than 60–$80 k_B T$ to guarantee 5 years of data storage. As the magnetic particle becomes smaller, thermal stability of the magnetic particle is degraded. One approach to resolve this impasse is to increase K_u to compensate for the decrease of V. However, the larger K_u causes the larger magnetic field to switch the magnetization of the materials. Therefore, to write data on the higher-K_u magnetic media, a higher magnetic field is required. The highest magnetic field obtainable by a magnetic write head is known to be nearly saturated.

7 Nanopatterned Media for High-Density Storage

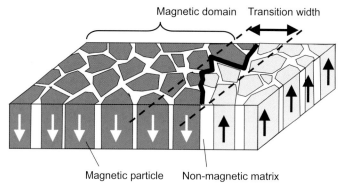

Figure 7.1 Schematic explanation of present-day magnetic media.

To overcome this situation, a hybrid recording method of optical and magnetic recording [1], schematically shown in Figure 7.2, is attracting much attention. In hybrid recording, data are written on high-K_u magnetic media by writing both with a magnetic field and an optical spot, which generate a heat spot to reduce switching of the magnetic field. In the hybrid recording, magnetic particles in the recording media can be miniaturized while maintaining sufficiently high $K_u V$. An optical device employing nanophotonics capable of generating near-field optical spots smaller than those attainable by diffraction of light is described in a later chapter. Achievement of hybrid recording HDD hinges on the development of a write head on which near-field optical devices are built close to the magnetic writer.

With regard to recording media for hybrid recording, various materials have been confirmed to have extremely high K_u values: synthetic magnetic multilayers of Co/Pd or Co/Pt, $L1_0$ ordered alloys of FePt, CoPt or FePd, and rare-earth transition-metal alloys of $SmCo_5$, TbFeCo, or $Nd_2Fe_{14}B$. In order to apply these high-K_u materials to extremely high-density magnetic recording media, it is necessary to obtain a fine granular structure within them.

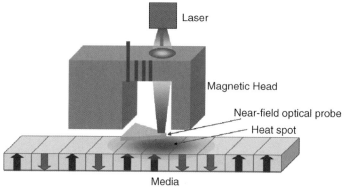

Figure 7.2 Schematic explanation of the hybrid recording method.

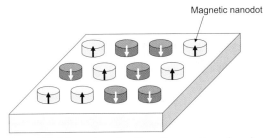

Figure 7.3 Schematic explanation of nanopatterned media.

Another approach to circumvent the superparamagnetic limit is the use of nanopatterned media [2, 3]. As shown in Figure 7.3, in nanopatterned media, recording layers consist of highly uniform magnetic dots that are regularly aligned on data tracks. With nanopatterned media, each magnetic dot can store 1 bit, and therefore, unlike in the case of granular media, 1 bit does not require hundreds of magnetic particles. Thus, the V of $K_u V$ in nanopatterned media is much higher than that in granular media.

The ultimate concept in magnetic recording is the combination of hybrid recording and nanopatterned media. For example, nanopatterned media with uniform $L1_0$ FePt particles with a diameter of 3 nm may achieve a density as high as 40 Tbpsi.

In this chapter, following a brief explanation of nanopatterned media, fabrication of $L1_0$ FePt nanopatterned media is described. As ultrafine lithography for high-density nanopatterned media, block-copolymer lithography was investigated.

7.2 Nanopatterned Media

In nanopatterned media, magnetic dots uniform in size and regularly aligned with uniform spacing are fabricated by lithography. Whereas, in conventional granular structure media, the SNR is determined by the magnetic particle size in magnetic recording layers, in nanopatterned media, the SNR is determined by the precision of the pattern size and the position of the magnetic dots. How to fabricate high-precision patterns has been a major issue in the context of efforts to realize nanopatterned media. With conventional perpendicular granular media, it has been thought that the growth of areal density saturates between 500 Gbpsi and 1 Tbpsi. Therefore, the density for nanopatterned media should be more than 1 Tbpsi. As schematically shown in Figure 7.4, to obtain this density, even with a bit aspect ratio of 1, which is the most favorable dimension for the patterning process, magnetic dots with a center-to-center distance of 25 nm should be fabricated. This pattern size remains challenging even with cutting-edge lithographic technology. Furthermore, a manufacturing process for achieving low cost and high throughput is critical. Although the idea of patterned media first appeared in the 1990s, no practical hard

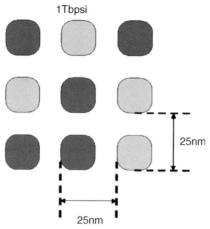

Figure 7.4 Example of alignment of magnetic dots at a density of 1 Tbpsi.

disk drives employing the technology have been commercialized. The main reason is that no practical lithographic technology has appeared.

7.2.1
Fabrication Process of Nanopatterned Media

Some of the fabrication processes for nanopatterned media are shown in Figure 7.5. With method A, an etching process is applied to continuous magnetic thin films to fabricate magnetic dots. Method A is advantageous in that it is easy to control the properties of magnetic thin films, but disadvantageous in view of the difficult of etching magnetic materials by a dry-etching process, which has been widely used for silicon materials. With method B, a magnetic recording layer is deposited on prepatterned substrates that have protruding dot arrays. The advantage of method B is that the process can be very simple. The main issue concerning method B is how to manufacture complicated film layer structures, such as the soft underlayers essential in conventional perpendicular media. In both methods, the mask pattern drawing process needs ultrahigh-resolution lithography.

In the semiconductor industry, lithography has been a leading technology. However, in the semiconductor industry, the near-term goal for lithography technology has been set at a center-to-center distance of around 50 nm, which is too large for the present-day mark size in hard disk drives. In conventional lithography, energy-beam irradiation causes dissociation of polymer chains or polymerization that can be resolved by certain developing processes. The resolution is determined by the diffraction limit of the energy beam as

$$r \propto \lambda/\mathrm{NA} \qquad (7.1)$$

where λ is the wavelength of energy beam, and NA the numerical aperture. To obtain a smaller pattern, light sources with shorter wavelengths have been developed, such as 436 nm/g line, 365 nm/i line, 248 nm/KrF excimer laser, and 193 nm/ArF excimer

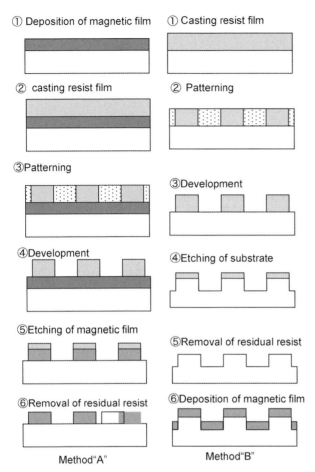

Figure 7.5 Examples of fabrication processes for nanopatterned media.

laser. Recently, use of extreme ultraviolet (EUV) light source whose wavelength is 13.6 nm is being investigated. According to the roadmap, lithography equipment with an EUV light source capable of generating a half-pitch of 22 nm is expected to be commercialized around 2015. In the semiconductor industry, the current development of lithography involves heavy investment in development of the light source, optical system and resist materials. The biggest advantage of the hard disk drive is low bit cost. The cutting-edge lithography such as EUV lithography is expected to be too expensive for production of HDD media. Novel lithography methods that preserve the bit-cost advantage of HDD are desired.

The most promising method of forming mask patterns for nanopatterned media is nanoimprint lithography (NIL) [2]. A schematic process flow of NIL is shown in Figure 7.6. In NIL, the pattern template made by another high-resolution lithography is pressed to resist thin film to transfer mask pattern on a substrate by deforming the resist film. The advantage of NIL is that the process is high

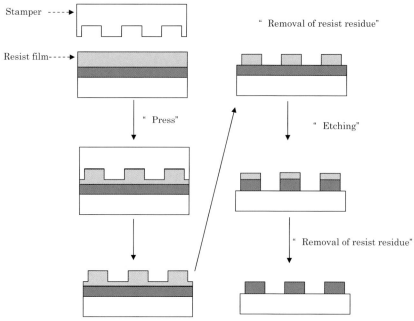

Figure 7.6 Nanoimprint lithography (NIL).

throughput and repetition of the NIL process many times with a template is possible once the template is made. Therefore, even if the template is made by expensive high-resolution lithography, the cost of high-resolution lithography can be negligible for a product.

Electron beam lithography (EBL) is widely used in the semiconductor industry to fabricate nanometer-scale devices. EBL has already been applied to the optical disk mastering process [4]. EBL is one of the promising candidate pattern drawing methods for nanopatterned media. Very recently, an application of EBL to the HDD mastering process has been demonstrated [5]. The possibility of high-precision patterning of HDD servo data has been confirmed.

The combination of EBL and NIL in which EBL is used to make NIL templates may be a practical solution for fabrication of nanopatterned media. It has been reported that 4-nm wide isolated lines are possible using EBL. However, a highly dense pattern, for example 20 nm center-to-center spacing, has not been resolved with a practical throughput. Further advances in beam size, beam brightness, reduction of proximity effect and highly sensitive high-resolution resists are required for EBL.

The other promising approaches for patterning of nanopatterned media involve self-assembling pattern formation. These approaches can lead to a higher pattern density unconstrained by the pace of advance in the performance of lithography equipment.

Porous alumite [6, 7], monolayers of synthesized nanoparticles [8], and block-copolymers [9] have already been applied to fabrication of nanopatterned media.

7.3
Block-Copolymer Lithography for Nanopatterned Media

7.3.1
Self-Assembled Phase Separation of Block-Copolymers

In this section, application of self-assembling periodic pattern formation of block-copolymers to nanopatterning is described. In Figure 7.7, a schematic structure of a block-copolymer and a schematic diagram of periodic pattern formation are shown. A block-copolymer has different polymer blocks covalently bonded to one another. Block-copolymers composed of two polymer blocks are called diblock-copolymers. Diblock-copolymers are widely studied, since they are relatively easy to synthesize. Block-copolymers composed of three polymer blocks are called triblock-copolymers. The physical properties of block-copolymers appear to reflect the properties of each polymer block. For example, in polystyrene-b-polybutadiene-b-polystyrenes, called SBS rubber, which are widely used as thermoplastic elastomers, polystyrene blocks exert high structural stiffness and polybutadienes exert high elasticity.

It is well known that block-copolymers show various patterns of phase separations depending on the polymer species or composition of polymer blocks. Because of repulsive interaction between different polymer blocks, the two different polymer blocks are immiscible and tend to separate from each other. In a polymer blends system, different polymers tend to form separated domains of relatively large size. On the other hand, in block-copolymers in which different polymers are combined covalently, phase separation occurs in smaller dimensions corresponding to the length of polymer chains, since different polymer blocks are restricted by the covalent bonds. The phase separation in a polymer blend system is called macrophase separation, and that in block-copolymers is generally called microphase separation.

Microphase separation behaviors have been widely investigated [10, 11]. The structure of microphase separation shows variations depending on the composition

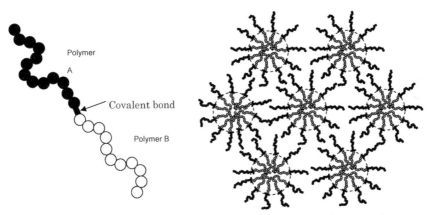

Figure 7.7 Block-copolymer and schematic diagram of phase separation of block-copolymers.

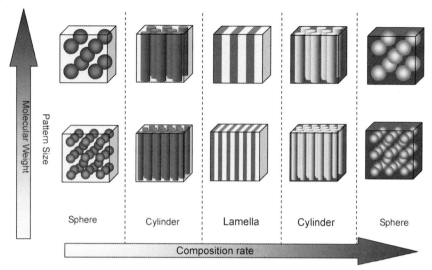

Figure 7.8 Schematic explanation of pattern variation of microphase separation.

ratio of polymer A and polymer B as shown in Figure 7.8. When the composition ratio is around 0.5, the diblock-copolymer tends to form a lamella structure. With the change to an asymmetric composition ratio, the structure varies from cylindrical to spherical. The size of these periodic structures is determined mainly by the molecular weight of the block-copolymer. Therefore, in order to miniaturize the pattern size, expensive equipment is not required and it is only necessary to synthesize the smaller block-copolymer molecules. The minimum pattern size is limited by the interaction strength of polymer blocks and by the molecular weight. In the case of the smaller block-copolymer molecules, the different polymer blocks tend to be more miscible. A center-to-center distance of as small as 10 nm phase separation with amphiphilic block-copolymers that show strong segregation has been reported [12]. This pattern size corresponds to the density of 5 Tdots/inch^2. Therefore, block-copolymer lithography is thought to be applicable as a fabrication method for nanometer-scale structures [13, 14] or devices [15], such as nanopatterned media [9, 16] beyond the density of 1 Tbpsi.

The process of pattern generation on a substrate is as follows. Block-copolymer films are deposited by spin coating with a dilute polymer solution. A block-copolymer film is in a disordered state, not in a thermal equilibrium state. A thermal annealing process at above the glass transition temperature of the polymer is needed to form an ordered phase-separated structure. It takes several hours to obtain an ordered pattern of high quality.

In order to apply the ordered periodic patterns to lithography, the two-dimensional patterns of block-copolymer thin films shown in Figure 7.9 are preferable. For example, monolayers of spheres or perpendicularly aligned cylinders are applicable to fabrication of regularly aligned dot patterns on substrates. To obtain monolayers of spheres, it is necessary to control the film thickness of block-copolymers. To obtain

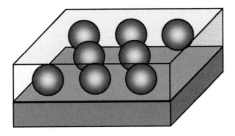

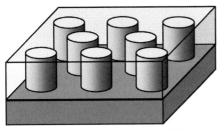

Figure 7.9 Two-dimensional patterns of block-copolymer phase separations.

perpendicularly aligned cylinders, it is necessary to control the film thickness or to control chemical properties of the surface on which the block-copolymers are coated [17].

7.3.2
Fabrication of Magnetic Nanodots by Block-Copolymer Lithography

Figure 7.10 shows a schematic explanation of the fabrication of magnetic nanodots by block-copolymer lithography. We adopted polystyrene(PS)-b- polymethylemethacrylate(PMMA) in view of the ease of transforming the phase-separated patterns to topographic patterns. PMMA can be removed by various methods, such as UV exposure, electron beam or oxygen plasma. PS-PMMA diblock-copolymers were solved in propylene glycol monomethyl ether acetate (PGMEA) with 1–3%(w/w). Solutions of PS-PMMA were spin-coated onto magnetic films and annealed at 180 °C in vacuum for 10 h to make ordered self-assembled dot arrays of PMMA. The PMMA domains are selectively removed by oxygen plasma etching to obtain hole arrays. Then, the holes are filled with spin-on-glass (SOG) by spin-coating with SOG solutions. The PS matrix is then removed by reactive ion etching (RIE) with oxygen gas. We obtain convex island structures. Topographic patterns of residual SOG dot arrays are transferred to the magnetic films by Ar-ion milling.

Results of FePt nanodot fabrication are reported. $L1_0$ FePt is a promising candidate recording material for optomagnetic hybrid recordings. NiTa/Cr/Pt/FePt films were deposited by DC magnetron sputtering methods. Substrates were heated during deposition using an infrared heater. No post-deposition annealing was performed. PS-b-PMMA (PS: 65 000 dalton, PMMA: 13 200 dalton) was used with no

7 Nanopatterned Media for High-Density Storage

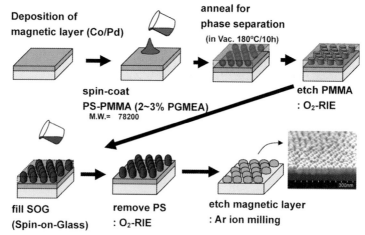

Figure 7.10 Schematic explanation of the fabrication method for nanopatterned media by block-copolymer lithography.

purification. This block-copolymer can form a spherical structure with a center-to-center distance of 30 nm.

Figure 7.11 shows a SEM image of a FePt dot array fabricated by ion milling for 90 s. The thickness of the FePt film was 20 nm. The average dot diameter was measured to be about 25 nm. Figure 7.12 shows a cross-sectional TEM image of the FePt dots. The diameter at the top and that at the bottom of the dots is about 20 nm and about 25 nm, respectively.

Figure 7.13 shows a high-resolution cross-sectional TEM image of ordered perpendicular FePt dots fabricated from FePt films. The thickness of the FePt films was 10 nm. It is clearly observed that the FePt dots were completely isolated by the

Figure 7.11 SEM image of FePt nanodots fabricated by block-copolymer lithography.

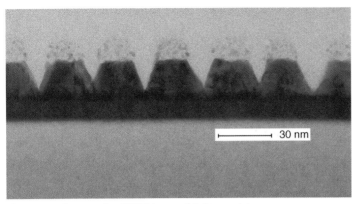

Figure 7.12 Cross-sectional TEM image of FePt nanodots.

ion-beam etching. The layer-by-layer crystal structures indicating (001) oriented $L1_0$ FePt were observed in dots. And the crystal structure was observed to be maintained even on the sidewalls. This shows the ion-beam etching causes low damage to the nanometer-scale magnetic dots. The thickness of a magnetic dead layer induced by ion-beam etching on NiFe film was reported to be 1.4 nm at 400 eV with an incident angle of 25° [18]. In the present case, the dead layer is thought to be much thinner than that, because the ion beam falls at much lower angle on a sidewall surface of the FePt dot.

The magnetic properties were measured by a vibrating sample magnetometer (VSM) with an applied field of ±20 kOe. Figure 7.14 shows magnetic hysteresis curves measured by VSM. Before etching, the coercive force was 4.5 kOe, and after etching, the H_c increased to 13 kOe. The squareness of the loops was 1.0. A low H_c, 700 Oe, was observed in the in-plane. The magnetization of the dots was not saturated

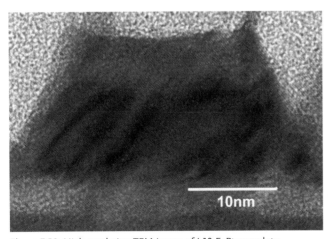

Figure 7.13 High-resolution TEM image of L10 FePt nanodots.

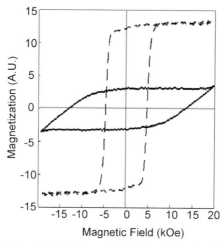

Figure 7.14 M–H loop of FePt film (dashed line) and dots (solid line).

because the magnetic field our VSM could generate was limited to 20 kOe. The observed large increase of the H_c was thought to be attributable to enhancement of shape anisotropy by formation of isolated dots. Huge H_c exceeding 100 kOe has been reported, which was expected from the coherent magnetization rotation for highly ordered L10 FePt particles with a diameter of about 20 nm [19]. The observed H_c in the present study was far smaller than the reported value. One possible explanation of the small H_c may be that the magnetization reversal for the present FePt dots was governed by nucleation of a reversed domain in a dot at a low magnetic field causing rapid domain-wall propagation. The same explanation was advanced for large patterned magnetic islands showing Stoner–Wohlfarth-like behavior in magnetization reversal [20, 21]. It was also reported that H_c for patterned FePt dots with $0.2 \times 0.2\,\mu m^2$ was increased by post-annealing. This was explained by the decrease in the number of nucleation sites due to post-annealing [22]. They also mentioned that degradation of the degree of crystal order may cause lowering of H_c. For the present result, the possibility of the existence of a similar degradation of ordered phase in the FePt dot cannot be negligible.

7.4
Control of Orientation of Self-Assembled Periodic Patterns of Block-Copolymers

In the microscopic view, the self-assembled periodic pattern of block-copolymers is close-packed ordered; however, in the macroscopic view, as shown in Figure 7.15, it has multidomains whose orientations of periodic patterns differ. In order to apply block-copolymer lithography to fabrication of nanopatterned media, control of the orientation of the periodic patterns is indispensable, since each dot should be accessible with a magnetic head on a rotating disk. In order to control the orientations of self-assembled periodic patterns of block-copolymers, approaches involving the

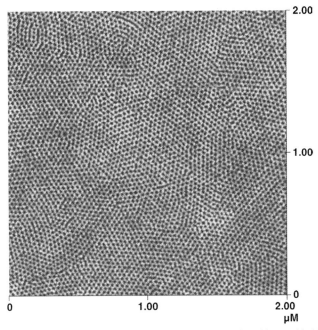

Figure 7.15 AFM image of a multidomain structure of a self-assembled block-copolymer film.

guiding of self-assembly by topographic patterns [23–25] or chemical patterns [26, 27] formed on substrates have been investigated. Considering that development of the existing top-down lithography appears to be reaching its limit, the combination of top-down and bottom-up lithography, which involves a self-assembling method, is becoming more important.

Figure 7.16 shows the outline of the process of orientation control of the periodic pattern formation of the block-copolymers' phase separations. On the topographic guide patterns, the block-copolymer films are confined within the grooves and the alignment of the self-assembled pattern follows the guide orientation automatically. For fabrication of guide patterns, NIL is cost effective [28]. Figure 7.17 shows AFM images of alignments of dot arrays of self-assembled PS-PMMA block-copolymer. To achieve defect-free periodic patterns of high quality, it is important to adjust the size of guide patterns to the built-in size of the self-assembled block-copolymers. We also investigated the effect of sectional shape of the guide patterns on the quality of the self-assembly. Figure 7.18 shows disordered dot arrays of self-assembled PS-PMMA block-copolymer on U-sectional shaped groove guides. U-shaped groove guides were fabricated by NIL on a photocurable polymer film and following etching by oxygen plasma treatment. In the case of loose confinement of block-copolymers, restriction of the self-assembly orientation by the guide patterns may not work well. It should be noted that, in addition to the precise control of the size of guide patterns, a high degree of lateral confinement is important for the fabrication of highly ordered periodic patterns of self-assembled block-copolymers.

160 | *7 Nanopatterned Media for High-Density Storage*

Multidomain structure of block copolymer

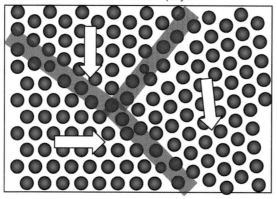

Periodic structure aligned by topographic guide pattern

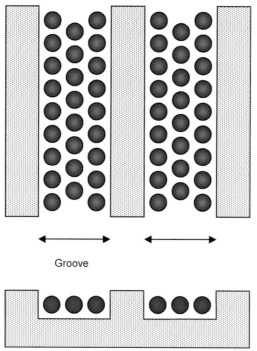

Groove

Figure 7.16 The outline of orientation control of the periodic pattern formation of the block-copolymers' phase separations.

We demonstrated fabrication of nanopatterned media by the guided block-copolymer lithography. Figure 7.19 is a schematic depiction of the process flow for guided block-copolymer lithography. A Ni master disk that has inverted line-and-groove guide patterns was fabricated by EBL with a rotating wafer stage, following Ni electroforming. The Ni master disk is pressed to a resist film on a FePt magnetic film

Groove width = 80nm

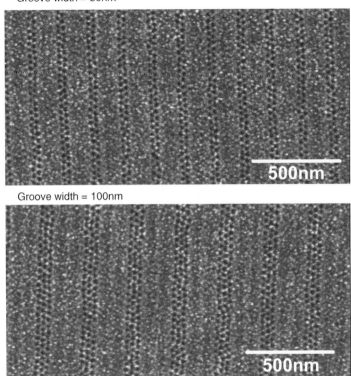

Groove width = 100nm

Figure 7.17 The groove-width dependence of 2D structures of guided self-assembly of PS-PMMA block-copolymers (PS: 65 000 dalton, PMMA: 13 200 dalton).

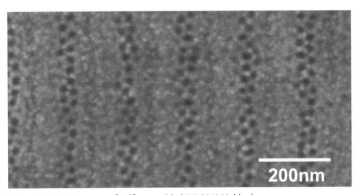

Figure 7.18 Dot arrays of self-assembled PS-PMMA block-copolymer on U-sectional-shaped grooves.

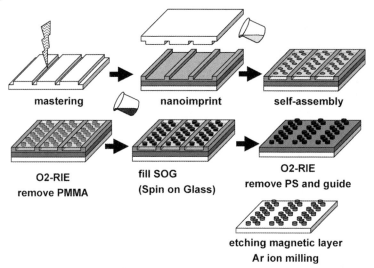

Figure 7.19 Schematic explanation of guided block-copolymer lithography.

to transfer the line-and-groove patterns. A PS-PMMA block-copolymer diluted solution is cast into the grooves, and then annealed to drive phase separation. The obtained PMMA dots are removed by dry etching of oxygen plasma. After this step, the process flow is the same as described above. The nanopatterned media of the obtained regularly aligned FePt dots fabricated using the guided self-assembling method is shown in Figure 7.20. 13 rows of FePt dots with 30 nm center-to-center distance and 20 nm height were observed. Figure 7.21 shows a cross-sectional TEM image of the FePt dot arrays sectioned orthogonally to the guide axis. In this nanopatterned media, a density of 500 Gdots/inch^2 was achieved. At present, a significant amount of defects exists in a periodic dot pattern. To achieve synchronous writing on a nanopatterned media, accuracy of the position of each dot is important. Methods of eliminating defects throughout 100 μm length along the guide grooves have yet to be developed.

7.5
Summary

The concept of merging hybrid recording and nanopatterned media is one of the promising approaches for realizing recording density of HDD beyond 10 Tbpsi. In this chapter, we described the fabrication of nanopatterned media with high-K_u magnetic recording material, L1$_0$ FePt, for the recording medium of hybrid recording. We apply block-copolymer lithography for fabrication of nanopatterned media. It is confirmed that the obtained FePt dot arrays have strong perpendicular anisotropy. Fabrication of aligned FePt dot arrays on circumferential lines was also demonstrated successfully using guided self-assembly of PS-PMMA diblock-copolymers by groove guide patterns.

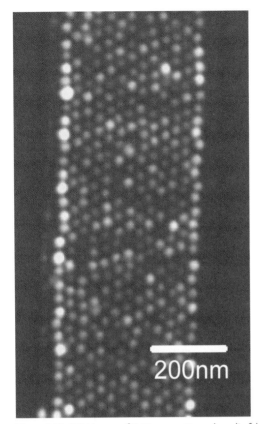

Figure 7.20 AFM image of FePt nanopatterned media fabricated by guided block-copolymer lithography.

Figure 7.21 Cross-sectional TEM image of FePt nanopatterned media fabricated by guided block-copolymer lithography.

Acknowledgements

The author would like to thank K. Naito, A. Kikitsu, N. Kihara, T. Maeda, Y. Yanagita, K. Kashiwagi, Y. Isowaki, M. Sakurai, Y. Kamata, K. Asakawa and T. Hiraoka of the Corporate Research & Development Center, Toshiba Corporation, for support and helpful discussion. This paper is part of the Terabyte Optical Storage Technology project funded by the Ministry of Economy, Trade and Industry of Japan (METI), which was consigned to the Optoelectronic Industry and Technology Development Association (OITDA) in 2002 and subsequently consigned to the New Energy and Industrial Technology Development Organization (NEDO) from 2003 onward.

References

1 Saga, H., Nemoto, H., Sukeda, H. and Takahashi, M. (1999) *Japanese Journal of Applied Physics*, **38**, 1839.
2 Chou, S.Y., Woi, M.S., Krauss, P.R. and Fisher, P.B. (1994) *Journal of Applied Physics*, **76**, 6673.
3 White, R.L., New, R.M.H. and Pease, R.F.W. (1997) *IEEE Transactions on Magnetics*, **33**, 990.
4 Wada, Y., Katsumura, M., Kojima, Y., Kitahara, H. and Iida, T. (2001) *Japanese Journal of Applied Physics*, **40**, 1653.
5 Wada, Y., Tanaka, H., Kitahara, H., Ozawa, Y., Hokari, M., Nishida, T., Suzuki, T., Yamaoka, M. and Sugiura, S. (2008) *Japanese Journal of Applied Physics*, **47**, 6007.
6 Nielsch, K., Muller, F., Li, A.-P. and Gosele, U. (2000) *Advanced Materials*, **13**, 582.
7 Oshima, H., Kikuchi, H., Nakao, H., Morikawa, T., Matsumoto, K., Nishio, K., Masuda, H. and Itoh, K. (2005) *Japanese Journal of Applied Physics*, **44**, L1355.
8 Sun, S., Murray, C.B., Weller, D., Folks, L. and Moser, A. (2000) *Science*, **287**, 1989.
9 Hawker, C.J. and Russell, T.P. (2005) *MRS Bulletin*, **30**, 952.
10 Bates, F.S. and Fredrickson, G.H. (1999) *Physics Today*, **52**, 32.
11 Hamley, I.W. (1998) *The Physics of Block-Copolymers*, Oxford University Press, New York.
12 Tian, Y., Watanabe, K., Kong, X., Abe, J. and Iyoda, T. (2002) *Macromolecules*, **35**, 3739.
13 Mansky, P., Chaikin, P. and Thomas, E.L. (1995) *Journal of Materials Science*, **30**, 1987.
14 Park, M., Harrison, C., Chaikin, P.M., Register, D.H. and Adamson (1997) *Science*, **276**, 1401.
15 Black, C.T., Ruiz, R., Breyta, G., Cheng, J.Y., Colburn, M.E., Guarini, K.W., Kim, H.-C. and Zhang, Y. (2007) *IBM Journal of Research and Development*, **51**, 605.
16 Hieda, H., Yanagita, Y., Kikitsu, A., Maeda, T. and Naito, K. (2006) *Journal of Photopolymer Science and Technology*, **19**, 425.
17 Thurn-Albrecht, T., Steiner, R., deRouchey, J., Stafford, C.M., Huang, E., Bal, M., Tuominen, M.T., Hawker, C.J. and Russell, T.P. (2000) *Advanced Materials*, **12**, 787.
18 Si, W., Williams, K., Campo, M., Mao, M., Devasahayam, A. and Lee, C.-L. (2005) *Journal of Applied Physics*, **97**, 10, N901–1.
19 Shima, T., Takahashi, K., Takahashi, Y.K. and Hono, K. (2004) *Applied Physics Letters*, **85**, 2571.
20 Dittrich, R., Hu, G., Schrefl, T., Thomson, T., Suess, D., Terris, B.D. and Fidler, J. (2005) *Journal of Applied Physics*, **97**, 10, J705–1.
21 Kikuchi, N., Murillo, R., Lodder, J.C., Mitsuzuka, K. and Shimatsu, T. (2005) *IEEE Transactions on Magnetics*, **41**.
22 Seki, T., Shima, T., Yakushiji, K., Takahashi, K., Li, G.Q. and Ishio, Shunji (2005) *IEEE Transactions on Magnetics*, **41**, 3604.
23 Segalman, R.A., Yokoyama, H. and Kramer, E.J. (2001) *Advanced Materials*, **13**, 1152.
24 Cheng, J.Y., Ross, C.A., Smith, H.I. and Thomas, E.L. (2006) *Advanced Materials*, **18**, 2505.

25 Yang, X.-M., Xiao, S., Liu, C., Pelhos, K. and Minor, K. (2004) *Journal of Vacuum Science and Technology*, **B22**, 3331.
26 Rockford, L., Liu, Y., Mansky, P., Russell, T.P., Yoon, M. and Mochrie, S.G.J. (1999) *Physical Review Letters*, **82**, 2602.
27 Yang, X.-M., Peters, R.D., Nealey, P.F., Solak, H.H. and Cerrina, F. (2000) *Macromolecules*, **33**, 9575.
28 Naito, K., Hieda, H., Sakurai, M., Kamata, Y. and Asakawa, K. (2002) *IEEE Transactions on Magnetics*, **38**, 1949.

8
Nanophotonics Recording Device for High-Density Storage
Tetsuya Nishida, Takuya Matsumoto, and Fumiko Akagi

8.1
Introduction

In this chapter we describe a nanophotonic recording device with a near-field optical probe for a high-density storage system with a data density of up to 1 Tbit/in^2. When the areal density of magnetic recording approaches 1 Tb/inch^2, thermal instability due to the superparamagnetic limits of magnetic materials has to be overcome [1, 2]. In a conventional granular medium, each data bit is made up of many tiny magnetic grains. Maintaining sufficient signal-to-noise ratio at high areal densities requires reduction of the grain volume, which causes thermal fluctuation of the magnetic domains. To suppress this thermal instability, we need to increase the coercivity of the medium. However, conventional magnetic heads cannot generate magnetic fields strong enough to switch the magnetization. One solution is to use thermally assisted magnetic recording (TAR) [3], in which the medium is heated by a laser during the writing process to reduce its coercivity. This enables us to use media with high coercivity. Another approach is to use bit-patterned media (BPM), in which data bits written on lithographically patterned magnetic islands behave as single magnetic domains [4]. We can push up the density limit because the volume of the magnetic domain increases. Hybrid recording using a combination of TAR and BPM, i.e. writing data with an irradiating laser on magnetic islands made of material with high coercivity [5], should be able to achieve an areal density of over 100 Tb/inch^2.

Hybrid recording requires TAR with a heated area as small as a data bit to produce a coercivity of less than the recording magnetic head's field. To obtain such a small heated area, the size of the optical spot should be less than a few tens of nanometers to prevent adjacent tracks from being erased. It is necessary to use an optical near-field [6] to generate such small optical spots that are beyond the diffraction limit of light. In order to generate the optical near-field efficiently, several kinds of highly efficient optical near-field generators, including a multitapered probe [7, 8], a silicon pyramidal probe [9], a c-aperture [10], and an aperture with circular grooves [11], have been studied. An antenna-type near-field generator with a beaked metallic plate

Nanophotonics and Nanofabrication. Edited by Motoichi Ohtsu
Copyright © 2009 WILEY-VCH Verlag GmbH & Co. KGaA, Weinheim
ISBN: 978-3-527-32121-6

(called a 'nanobeak') [12, 13] is particularly effective. Marks with 40-nm diameters were written on a GeSbTe phase-change recording medium by a head with a nanobeak. Hybrid recording also requires BPM with magnetic single-domain islands less than a few tens of nanometers—as small as a data bit. The BPM has to be covered with such small magnetic dots. A self-assembling method with diblock-copolymers is suitable for obtaining BPM with arrays of the small dots [14, 15]. To make a practical hybrid recording system, it is necessary to fabricate circularly aligned Co/Pd BPM using a guide drawn with a rotary-stage e-beam tool [16].

In this chapter we describe in detail hybrid recording in which a 'nanobeak' near-field optical probe was used to write marks on the BPM.

8.2
Thermally Assisted Magnetic Recording Simulation

Thermally assisted magnetic recording (TAR) on a small heated area with a near-field optical head has two advantages in high-density magnetic recording. One is reducing the effect of the magnetic switching field, and the other is increasing the effect of magnetic transition steepness during recording. When the temperature dependence of the magnetic switching field is dH_{co}/dT, a head field H_w for writing is expressed by the equation

$$H_w = H_{w0} + \frac{dH_{c0}}{dT}\Delta T_{max}$$

Here, H_{w0} is the head field without thermal assistance, and ΔT_{max} is the rising temperature at the heated center. When the head-field gradient is dH_{head}/dx, the effective head-field gradient dH_{eff}/dx is expressed by the equation

$$\frac{dH_{eff}}{dx} = \frac{dH_{head}}{dx} + \frac{-dH_{c0}}{dT}\frac{dT}{dx}$$

These two effects are shown together in Figure 8.1. On the basis of the effects, TAR on a small heated area was simulated. Magnetization (M_{rec}/M_s) during hybrid recording was simulated to determine the advantage it has over conventional perpendicular magnetic recording using a simplified model method [17], where M_{rec} is the recorded magnetization and M_s is the saturation magnetization. This method is based on five head-medium models: (1) bar magnets for head-field model, (2) switching integration force model, in which magnetization reverses when the time integration of the head-field equals switching energy, (3) Gaussian distribution for a thermal profile, (4) pulse waves and continuous heat radiation as a function of time, and (5) linear function for temperature dependences of K_u and M_s, where K_u is the anisotropy. The thermal stability of a magnetic domain depends on K_β ($K_u V/kT$), where V is particle volume, T is temperature, and k is the Boltzmann constant. Because a stable domain requires a K_β value of 60 or more [1], small grains require a high K_u value and a high recording-head field (H_h). Using a granular medium with 8-nm grains and a K_u value of 2.5×10^5 J/m³ ($K_\beta = 61$), the performance of

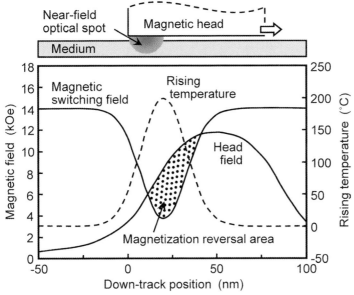

Figure 8.1 Conceptual diagram of thermally assisted magnetic recording. (Relationship between magnetic switching field and head field under near-field optical spot.)

conventional perpendicular recording with H_h of 14 kOe was evaluated, as shown in Figure 8.2(a). Using a granular medium with 4.5-nm grains and a K_u value of 8×10^5 J/m^3 ($K_\beta = 62$), the performance of hybrid recording with H_h of 14 kOe was evaluated, as shown in Figure 8.2(b). Domains were not formed during conventional recording because of an insufficient head field; however, magnetization modulation

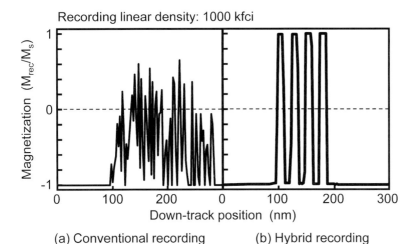

(a) Conventional recording (b) Hybrid recording

Figure 8.2 Comparison of recording simulations between conventional and thermally assisted magnetic recording.

was very clear during hybrid recording. Hybrid recording is effective even when the grains are as small as 5 nm or less [18].

8.3
The 'Nanobeak,' a Near-Field Optical Probe

Figure 8.3 shows a schematic of the near-field optical head used for the experiment. The head consisted of a triangular gold (Au) plate 'nanobeak' embedded in the surface of a quartz slide. When light polarized in the x-direction is introduced into the metallic plate, localized plasmons in the metal are excited and vibrating charges concentrate at the apex. The concentrated charges generate a strong optical near-field near the apex. The plate has a beaked apex and a recess near the edge opposite it. The beaked apex can generate a smaller optical spot than a plate without a beaked apex because charges are concentrated in a smaller area. To fabricate the plate, a triangular depression was formed on the surface of a quartz slide using electron-beam lithography, and gold film was deposited in the depression with a vacuum evaporation unit. The surface of the gold film was etched with an ion-milling machine to make the recess. Figure 8.4 shows a scanning electron microscope (SEM) image of the prototype nanobeak. Thickness (t), recess depth (d), length (l), and apex angle (θ) were 55 nm, 15 nm, 100 nm, 55 degrees, respectively. The apex radius of the fabricated plate was about 13 nm. An intensity distribution was calculated in the optical near-field generated by the metallic plate with the above dimensions by the finite-difference time-domain (FDTD) method. Figure 8.5 shows how the calculated intensity distribution looked on a plane 8 nm from the metallic plate.

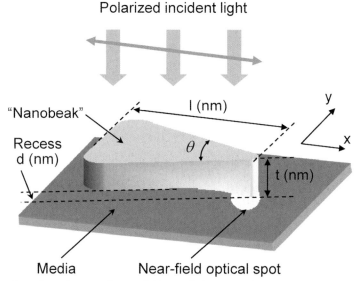

Figure 8.3 Schematic of a near-field optical plasmon probe 'nanobeak'.

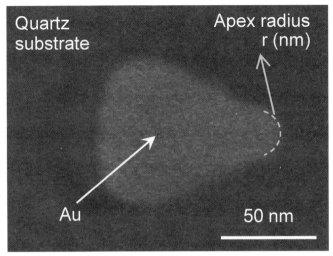

Figure 8.4 SEM image of the prototyped 'nanobeak'.

For this calculation, it was assumed that a 20-nm thick Co recording film was placed near the metallic plate and the spacing between the plate and the recording film was 8 nm. A plane wave with a wavelength of 780 nm was introduced into the plate. As shown, the peak intensity was 1400 times that of the incident light, and the width of

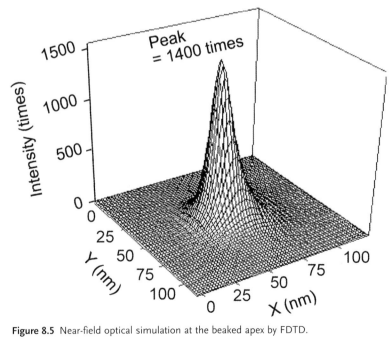

Figure 8.5 Near-field optical simulation at the beaked apex by FDTD.

the optical spot measured at half of the maximum was 16 nm in the x direction and 20 nm in the y direction [18, 19].

8.4
Bit-Patterned Medium with Magnetic Nanodots

The bit-patterned medium was fabricated by cutting a magnetic thin film into magnetic nanodots using self-assembled block-copolymers [20]. The fabrication process of the magnetic nanodot was described in Chapter 7 in detail. Figure 8.6(a) shows the structure of the medium. First, 5-nm thick Ti and 10-nm thick Pt layers were deposited on a quartz glass substrate and a 20-nm thick Co (0.3 nm)/Pd (0.7 nm) multilayer was deposited on them with a sputtering machine. Polystyrene–polymethylmethacrylate (PMMA) block-copolymer solution was then coated on the magnetic film and annealed to generate a self-assembled dot array of PMMA. The Co/Pd film was etched by ion-milling using the PMMA dots as an etching mask. After the etching, carbon was sputter deposited to fill the space between dots to make the surface flat. Figure 8.6(b) shows an atomic force microscope (AFM) image of the Co/Pd dots observed before carbon was deposited between them. The diameter of each dot was around 25 nm and the pitch was 30 nm, corresponding to a recording density of 830 Gbit/in^2. After applying an almost saturated magnetic field to the medium, the surface of the medium was observed using magnetic force microscopy (MFM) and AFM, as shown in Figures 8.7(a) and (b), respectively. A single magnetic domain with a resolution almost the same as the dot was clearly visible as a 30-nm diameter bright dot in the MFM image. The anisotropy of the Co/Pd multilayer film was 9.2×10^5 J/m^3. The coercivity of the BPM was 9.5 kOe. The Curie temperature

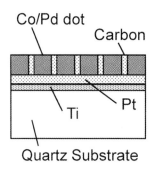

(a) Medium structure

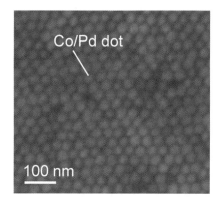

(b) AFM image of surface before carbon deposition

Figure 8.6 Co/Pd bit patterned medium.

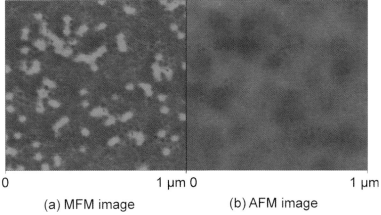

(a) MFM image (b) AFM image

Figure 8.7 Surface roughness and magnetic domain of Co/Pd BPM before initialization.

was 400 °C [21]. These magnetic properties indicate the Co/Pd BPM is suitable for hybrid recording [18].

To investigate the practical use of BPM, a circularly aligned 30-nm period Co/Pd-multilayer BPM was experimentally fabricated by arranging diblock-copolymers in guiding grooves. The 90-nm wide grooves were formed with a 30-nm diameter and ±30 nm wobbling e-beam using a rotary-stage e-beam mastering tool. A prototype of the circular 3-line aligned BPM fabricated with artificially assisted self-assembling (AASA) is shown in Figure 8.8 [18].

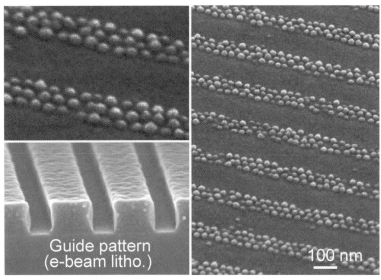

Figure 8.8 SEM image of a prototype of the circular 3-line aligned BPM using AASA for practical use.

8.5
Hybrid Recording Experiment

Using the Co/Pd BPM and the near-field optical head with the nanobeak, static recording was used to investigate hybrid recording characteristics. The quartz glass slide was placed on the medium and the medium was scanned with a piezo stage. Laser light with a wavelength of 785 nm was focused by an objective lens with a numerical aperture (NA) of 0.8 and introduced to the nanobeak. A magnetic field was applied to the medium with a bar magnet placed on the opposite side of the slide [18, 22].

Before the recording experiment, the BPM was uniformly initialized in a single perpendicular magnetic direction. Figure 8.9 shows MFM images of marks written on the medium with a mark pitch of 120 nm in the horizontal direction and 1 μm in the vertical direction. For this writing, the power and pulse width of the laser were, respectively, 8.5 mW and 25 ns, and the applied magnetic field was 1.8 kOe. As shown in the figure, the magnetizations of the dots were selectively reversed, and the minimum size of the reversed regions in the MFM image was around 30 nm (arrow A).

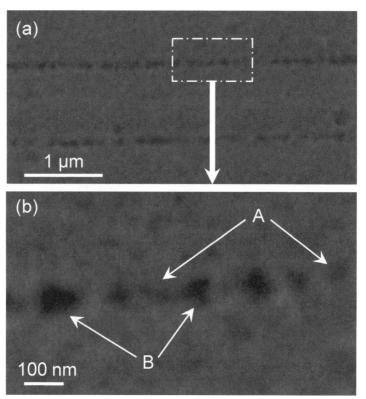

Figure 8.9 (a) MFM images of medium after writing marks. (b) Magnified view of the region surrounded by a dotted line in (a).

It is obvious that the size of a magnetic domain in the MFM image appears bigger than its actual size because of the resolution limit of the MFM, and the size of the regions that are magnetically reversed would be larger than 60 nm if two dots were magnetically reversed. Therefore, the magnetically reversed areas correspond to single dots. In the MFM image, larger marks are also observed (arrow B). In the experimental medium, the magnetic nanodots were not perfectly aligned to the direction of the sample scanning. It is expected that in these regions, the optical spot was not at the center of the dot, and several dots were heated simultaneously, which resulted in magnetization reversal of several dots. This was consistent with recording an isolated magnetic single dot on the Co/Pd BPM with a period of 30 nm [22].

8.6
Near-Field Optical Efficiency in Hybrid Recording

First, the near-field optical efficiency of the nanobeak in hybrid recording was calculated. Here, the efficiency (η) is defined by the equation

$$\eta \equiv \frac{Q_{abs}}{Q_{in}} = \frac{P_{abs}\Delta t}{P_{in}\Delta t} = \frac{P_{abs}}{P_{in}} = \frac{\int p_{abs} dV}{P_{in}}$$

Here, Q_{abs} is the absorbed energy in the whole medium, P_{abs} is the absorbed power in the whole medium, Q_{in} is the energy of the incident light, P_{in} is the power of the incident light, Δt is the pulse width of the irradiated light, p_{abs} is the absorbed power density in the medium, and V is the volume where the optical near-field is absorbed in the medium. When the irradiated light was focused with an objective lens with NA of 0.8, the efficiency calculated from the distribution for the nanobeak was 14% [18].

Next, the near-field optical efficiency of the nanobeak in hybrid recording was evaluated based on writing experiment results. Here, the writing temperature was estimated from the temperature dependence of the medium's coercivity. The coercivity (H_c) depends on the pulse width of light (t), and their relation is expressed by Sharrock's formula:

$$H_c(t) = H_{co}\left(1 - \left(\frac{\ln(t) + \ln(f_0/\ln 2)}{K_u V/kT}\right)^C\right)$$

Here, V and k are the volume of the dot and the Boltzmann constant. Constants f_0 and C are 1×10^{11} Hz and 0.7, respectively [23]. In the case of BPM, $H_{co} = \sim 2K_u/M_s$, where K_u and M_s are the anisotropic energy and saturation magnetization (M_s). The circular dots in Figure 8.10(a) represent the temperature dependence of M_s measured with a vibrating sample magnetometer, the solid line represents the Brillouin function fitted to the measured values, and the dotted line represents the temperature dependence of K_u calculated assuming $K_u(T)/M_s(T) = [K_u(0)/M_s(0)]^3$ [24]. In order to determine the absolute value of K_u, the time dependence of $H_c(t)$ was measured at room temperature by changing the scan times of the magneto-optical Kerr effect measurement (circular dots in Figure 8.10(b)) and fitting Sharrock's formula to them.

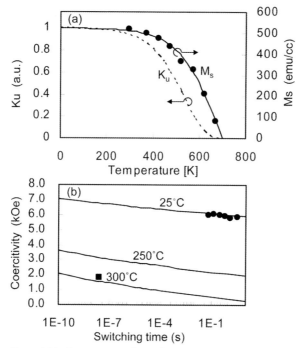

Figure 8.10 Characteristic of Co/Pd patterned media:
(a) Temperature dependence of saturation magnetization (M_s) and anisotropic energy (K_u). (b) Dependence of coercivity on temperature and switching time. In (a).

The obtained K_u at room temperature was 1.9×10^6 erg/cc. The solid lines in Figure 8.10(b) represent H_c at temperatures from 25 to 300 °C calculated using Sharrock's formula. The square dot in Figure 8.10(b) represents the writing condition in the experiment, and it is near the 300 °C line. Therefore, it was concluded that the writing temperature was about 300 °C [22].

The light-utilization efficiency of the head was roughly estimated from the experimental results. The efficiency was defined as the ratio between absorbed power in the medium and incident light power. In order to reverse the magnetization of a single dot, the medium temperature had to be elevated to over 300 °C, and the width of the area where the temperature is higher than 300 °C had to be less than 40 nm (diameter of the dot + width of carbon around the dot). The power required to generate such a temperature distribution was calculated by solving the thermal diffusion equation. It was compared with the input laser power in the experiment. For calculating the temperature distribution, it was assumed that a 20-nm thick Co/Pd recording layer with 5-nm thick Ti and 10-nm thick Pt underlayers was heated by a heat source with a Gaussian profile. The thermal conductivity of the Co/Pd layer, the width of the heat source, and the heating time were assumed to be 30 W/m K, 20 nm, and 25 ns, respectively. When the power of the heat source was adjusted so that the diameter of the circular contour of 300 °C was 40 nm, the power of the heat source was 0.5 mW. In the experiment, the input power was 8.5 mW. Therefore, the light-utilization efficiency was about 5%. Note that in the medium used for the experiments, the magnetic nanodots were surrounded by carbon sputtered using a target with a thermal conductivity of about 100 W/m K. However, it was presumed that the effective thermal conductivity of the deposited carbon was near the value of Co/Pd because the thermal conductivity of a sputtered film is normally smaller than its conductivity in the bulk, and thermal resistance at the boundary of Co/Pd dots suppresses thermal diffusion. Therefore, it was assumed that the recording layer was a Co/Pd film in our calculation. Modeling that takes into consideration the thermal impact of the material surrounding the dots will be needed for accurate calculation of the temperature distribution [22].

8.7 Summary

We introduced a hybrid recording method that recorded on a single dot on a 20-nm diameter nanodot Co/Pd-multilayer BPM using a near-field optical head with an antenna-type plasmon probe, a 'nanobeak'. The nanobeak, which uses a triangular metallic plate with a three-dimensionally tapered apex, can generate a highly efficient optical spot. The size of the optical spot generated at the apex is simulated by the FDTD method to be as small as 16×20 nm. The Co/Pd BPM consisted of a Co (0.3 nm)/Pd (0.7 nm) multilayer of 20-nm thick magnetic nanodots each with a diameter of 20 nm and a period of 30 nm, corresponding to a recording density of 830 Gb/in^2. It was confirmed that a single magnetic dot was individually recorded on the Co/Pd BPM. The light-utilization efficiency, defined as the ratio between the integration of the absorbed power in the medium and the incident light power, was estimated from the experimental results and thermal modeling to be about 5%. Hybrid recording simulation results suggest that hybrid recording with BPM can achieve stable, significant data-density recording, even when the dot size in the BPM is less than 5 nm.

Magnetic recording using a flying head with a waveguide was recently confirmed when a laser was induced through the waveguide [25]. We are looking forward to meeting the requirements for a data-storage device with Tbit/in^2 class or higher data density using the nanophotonic recording technology we have described here.

Acknowledgments

This work was performed as part of the Terabyte Optical Storage Technology Project (2002–2006 FY) of the Optoelectronic Industry and Technology Development Association (OITDA). OITDA contracted with the New Energy and Industrial Technology Development Organization (NEDO) to undertake the project based on funds provided from the Ministry of Economy, Trade and Industry of Japan (METI).

References

1 Charap, S.H., Lu, R.-L. and He, Y. (1997) *IEEE Transactions on Magnetics*, **33**, 978.
2 Hosoe, Y., Tamai, I., Tanahashi, K., Yamamoto, T., Kanbe, T. and Yajima, Y. (1997) *IEEE Transactions on Magnetics*, **33**, 3028.
3 Saga, H., Nemoto, H., Sukeda, H. and Takahashi, M. (1999) *Japanese Journal of Applied Physics*, **38**, 1839.
4 Chou, S.Y., Wei, M.S., Krauss, P.R. and Fisher, P.B. (1994) *Journal of Applied Physics*, **76**, 6673.
5 McDaniel, T.W. (2005) *Journal of Physics: Condensed Matter*, **17**, R315.
6 Ohtsu M. (ed.) (1998) *Near-field Nano/Atom Optics and Technology*, Springer, Tokyo.
7 Saiki, T., Mononobe, S., Ohtsu, M., Saito, N. and Kusano, J. (1996) *Applied Physics Letters*, **68**, 2612.
8 Yatsui, T., Kourogi, M. and Ohtsu, M. (1998) *Applied Physics Letters*, **73**, 2090.
9 Yatsui, T., Kourogi, M., Tsutsui, K., Takahashi, J. and Ohtsu, M. (2000) *Optics Letters*, **25**, 1279.
10 Shi, X., Hesselink, L. and Thornton, R.L. (2003) *Optics Letters*, **28**, 1320.
11 Thio, T., Pellerin, K.M., Linke, R.A., Lezec, H.J. and Ebbesen, T.W. (2001) *Optics Letters*, **26**, 1972.
12 Matsumoto, T., Shimano, T., Saga, H., Sukeda, H. and Kiguchi, M. (2004) *Journal of Applied Physics*, **95**, 3901.
13 Matsumoto, T., Anzai, Y., Shintani, T., Nakamura, K. and Nishida, T. (2006) *Optics Letters*, **31**, 259.
14 Naito, K., Hieda, H., Sakurai, M., Kamata, Y. and Asakawa, K. (2002) *IEEE Transactions on Magnetics*, **38**, 1949.
15 Hieda, H., Yanagita, Y., Kikitsu, A., Maeda, T. and Naito, K. (2006) *Journal of Photopolymer Science and Technology*, **19**, 425.
16 Wada, Y., Tanaka, H., Kitahara, H., Ozawa, Y., Hokari, M., Nishida, T., Suzuki, T., Yamaoka, M. and Sugiura, S. (2008) *Japanese Journal of Applied Physics*, **47**, 6007.
17 Igarashi, M., Akagi, F., Nakamura, A., Ikekame, H., Takano, H. and Yoshida, K. (2000) *IEEE Transactions on Magnetics*, **36**, 154.
18 Nishida, T., Matsumoto, T., Akagi, F., Hieda, H., Kikitsu, A., Naito, K., Koda, T., Nishida, N., Hatano, H. and Hirata, M. (2007) *Journal of Nanophotonics*, **1**, 011597.
19 Matsumoto, T., Nakamura, K., Nishida, T., Hieda, H., Kikitsu, A., Naito, K. and Koda, T. (2007) We-H-04, International Symposium on Optical Memory.
20 Hieda, H., Kikitsu, A., Koda, T., Maeda, T., Yanagita, Y., Kihara, N. and Naito, K. (2007) Th-PP-03 International Symposium on Optical Memory.
21 Koda, T., Hieda, H., Naito, K., Kikitsu, A., Matsumoto, T., Nakamura, K., Nishida, T. and Awano, H. (2007) AC-02 52nd Magnetism and Magnetic Materials Conference.
22 Matsumoto, T., Nakamura, K., Nishida, T., Hieda, H., Kikitsu, A., Naito, K. and Koda, T. (2008) *Applied Physics Letters*, **93**, 031108.
23 Igarashi, M. and Sugita, Y. (2006) *IEEE Transactions on Magnetics*, **42**, 2399.
24 Chikazumi, S. (1987) *Physics of Ferromagnetism*, vol. **2** Shokabo, Tokyo, pp. 29–30 (in Japanese).
25 Gage, E.C., Challener, W.A., Gokemeijer, N.J., Ju, G.P., Lu, B., Pelhos, K., Peng, C.B., Rottmayer, R.E., Yang, X.M., Zhou, H., Rausch, T. and Seigler, M.A. (2007) We-H-05 International Symposium on Optical Memory.

9
X-ray Devices and the Possibility of Applying Nanophotonics
Masato Koike, Shinji Miyauchi, Kazuo Sano, and Takashi Imazono

9.1
Introduction

X-ray spectrometry is one of the most essential tools available for investigating the properties of materials. For example, X-ray emission spectroscopy is widely used to obtain information on electron orbital energies and on densities of states of the valence bands. X-ray absorption spectroscopy is also a powerful technique for studying unoccupied electronic states and atomic structures. Improvements in X-ray spectrometric instruments are the driving force behind X-ray spectrometry progress, and these activities are summarized in Refs [1, 2].

The advent of advanced soft X-ray sources, including third-generation synchrotron radiation, X-ray lasers, and high-power X-ray plasma sources, together with increasing demand for their critical applications in science and industry, have pushed forward improvements in the performance of dispersing elements, especially in the 1 to 2 keV region. However, the monochromator crystals with large lattice constants that are suitable for this region have the drawbacks of low resistance to heat load and sometimes suffer from deliquescence, or from edges that absorb undesirable constituent elements [3]. The fundamental advantages of diffraction gratings are their excellent durability against heat load by using substrates with high thermal conductivity and a low coefficient of expansion [4], and no edges that absorb undesirable constituent elements. However, for these diffraction gratings to surpass the performance of monochromator crystals, substantial improvements in both diffraction efficiency and resolving power are required.

Most soft X-ray diffraction gratings are fabricated using mechanical ruling and holographic methods [5, 6]. For laminar-type holographic gratings, improvements in surface roughness of the grating substrate has increased diffraction efficiency and, along with reducing stray light, extended their practicable photon energy range up to approximately 1 keV or more [7–9]. Recently, significant efforts have been made to improve diffraction efficiency in two directions: one is towards the challenge of fabricating blazed-type gratings with extremely shallow blaze angles [10, 11]; the

Nanophotonics and Nanofabrication. Edited by Motoichi Ohtsu
Copyright © 2009 WILEY-VCH Verlag GmbH & Co. KGaA, Weinheim
ISBN: 978-3-527-32121-6

other is towards developing laminar-type multilayer gratings those are fabricated either by depositing a soft X-ray multilayer on a grating [12–14], or by generating a groove structure on the surface of a multilayer deposited on the substrate [15].

In addition to diffraction efficiency, it is indispensable for diffraction gratings designed for use in the multiple keV region to improve their resolving power, which is markedly inferior to that of monochromator crystals, which have small lattice constants in the order of a nanometer. Resolving power is inversely proportional to the grating constant and also proportional to the spectral order. Therefore, to improve their resolving power it is crucial that we develop diffraction gratings that have significantly smaller grating constants (or larger groove densities).

An advanced lithography system [16], based on optical near-field technology [17, 18] has made it possible to fabricate gratings that have attractive features: a high groove density compared to the conventional gratings made using mechanical ruling and holographic methods; and a large ruled area, compared to gratings made using electron-beam writing technology. In this chapter we describe the design and fabrication of a multilayer grating with a high groove density of 7600 lines/mm that is optimized for the 1 to 2 keV region. We also observed the surface of the fabricated grating, compared its diffraction efficiency with that of a typical monochromator crystal, and analyzed the difference of the experimental and theoretical diffraction efficiencies by applying two types of Debye–Waller factors, in terms of rms roughness.

9.2
Design of the Multilayer Laminar-Type Grating

In this section we briefly summarize a kinematic theory of multilayer laminar-type gratings, which we used to find the wavelength and geometric conditions that simultaneously satisfy the diffraction conditions for the grating and multilayer structures, and provide a peak diffraction efficiency. Also, we describe the conditions for determining groove depth, and for the selecting of material pairs for the multilayer used in this work.

The schematic diagram shown in Figure 9.1 is of a laminar-type multilayer grating that has been deposited with multiple layers, and that is the reiteration of two kinds of layers on laminar-type grooves. In this figure, the origin of the coordinate system, O, is at the center of the grating and the x-axis is the grating normal at O. When the grating is illuminated at an incident angle, α, and the diffracted rays of wavelength λ are observed from an angle, β, the sign of angle α (or β) is taken as positive or negative according to whether the projection of the incident ray (or diffracted ray) onto the x–y plane lies in the first or in the fourth quadrant of the x–y plane. Additionally, D, d, a, and h are the grating constants, multilayer period length, land width, and groove depth, respectively.

The grating equation and the generalized Bragg condition are denoted as [19]:

$$m_G \lambda = D(\sin\alpha + \sin\beta) \tag{9.1}$$

and

$$m_M \lambda = d(R_\alpha \cos\alpha + R_\beta \cos\beta) \quad (9.2)$$

where

$$R_\eta = \sqrt{1-(2\delta-\delta^2)/\cos^2\eta}, \quad \eta = \alpha, \beta \quad (9.3)$$

m_G, m_M, and δ are the spectral order of the grating, the order of the generalized Bragg condition, and the composite average of the real deviation of the index of refraction from unity for the unit area indicated by the dotted line in Figure 9.1. The incidence angle, α, and diffraction angle, β, which gives a peak diffraction efficiency at a wavelength of λ, are both determined by solving (9.1) and (9.2) simultaneously. In the design stage, it is common practice to use the tabulated refractive index [20, 21] of the two multilayer materials to calculate the value of δ.

Generally, the zeroth-order efficiency of a shallow laminar-type grating can be suppressed by controlling the groove depth. This feature is also utilized to make a simple measurement of groove depth. If destructive interference between the reflected wavefronts from the land and groove facets occurs for the zeroth-order light, the relation between the wavelength and incidence angle is expressed by

$$\lambda = 4h\cos\alpha \quad (9.4)$$

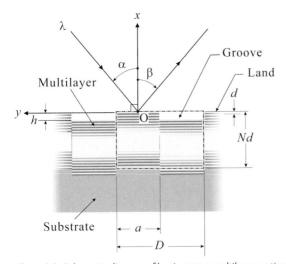

Figure 9.1 Schematic diagram of laminar-type multilayer grating. The area involved to derive the composite average of the real deviation of the index of refraction from unity, δ is indicated with a dashed line.

Another consideration is that a laminar-type grating may enhance the intensity of a spectral-order light by matching the phase of the radiations from the grooves and lands constructively. This condition is denoted by

$$m_G \lambda = 2h(\cos\alpha + \cos\beta) \tag{9.5}$$

where m_G is the diffraction spectral order that should be an odd number [22].

The selection of the materials pair for the grating's multilayer should be decided using a systematic materials survey. Figure 9.2 shows a typical example of the reflectivity at the position of Bragg peaks for multilayer mirrors. These mirrors are composed of material pairs of Mo/SiO$_2$ [15], Co/SiO$_2$ [14], Mo/C, and Pt/C [1, 12] in the region of 1 to 2 keV. The assumed multilayer period length, d, and the number of periods, N, are 5.5 nm and 30 periods (60 layers), respectively. We assumed the ratio of the thickness of the heavy material, d_1, to the period length of the multilayer, d, $\Gamma = d_1/d$, was to be 0.4 and that the tabulated refractive indices [20, 21] were used. The incidence angle, α, was varied according to the wavelength to satisfy the Bragg condition. As a result it was easily found that the Mo/C and Mo/SiO$_2$ multilayers provide high reflectivity in the wavelength range of 0.6 nm to 1.2 nm, which corresponds to an energy region of approximately 1 keV to approximately 2 keV. We decided to use an Mo/SiO$_2$ multilayer, since SiO$_2$ is known as a physically stable material [23] and the sputtering rate of carbon is exceptionally slow, approximately two orders of magnitude slower than SiO$_2$, by an ion-beam sputtering method we employed.

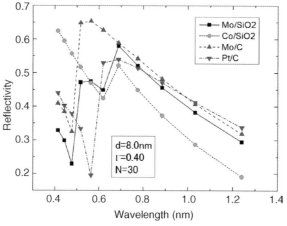

Figure 9.2 First-order Bragg reflectivities calculated for various multilayer mirrors. A 5.50-nm period length and 30 periods were assumed for the multilayers.

9.3
Specification of the Multilayer Laminar-Type Grating

To evaluate the diffraction efficiency of the various grating models, specifications of the multilayer lamina-type gratings we fabricated were determined using a numerical simulation. The assumed conditions were a groove density of $1/D = 7600$ lines/mm, diffraction orders of $m_G = m_M = +1$, a Mo thickness ratio of d_1, a multilayer period length of $\Gamma = 0.4$, a number of periods of $N = 30$, and a wavelength region of 0.6 nm to 1.2 nm. The incidence angle was restricted to be <88.5°, to accept a practical incoming amount of flux by the grating. The final design parameters were $h = 3$ nm, $a/D = 0.4$, and $d = 5.5$ nm. Figure 9.3 shows diagrams of (a) the incidence and diffraction angles calculated by solving (9.1) and (9.2), and (b) groove depths calculated by (9.4) and (9.5) versus the wavelengths for the multilayer period lengths of 5.0, 5.5, and 6.0 nm. We found that the designed groove depth of 3 nm is ultimately a compromise value, from the groove depths obtained by (9.4) and (9.5).

9.4
Fabrication of Multilayer Laminar-Type Gratings

Our multilayer grating was fabricated by depositing an Mo/SiO$_2$ multilayer onto the surface of a laminar-type master grating. This laminar-type master grating was fabricated by Tadashi Kawazoe and Motoichi Ohstu at the University of Tokyo, using state-of-the-art optical near-field lithography technology [17, 18]. A stable, computer-controlled prototype for these commercial products was developed in collaboration with T.K., M.O. and Canon Inc. [16, 24], resulting in a desktop-sized machine with a

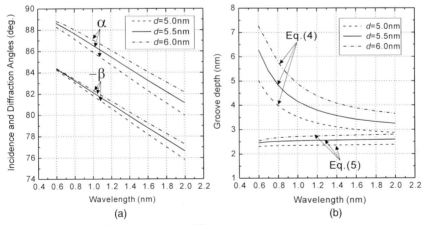

Figure 9.3 Diagrams of (a) incidence and diffraction angles calculated by solving (9.1) and (9.2) assuming $h = 3$ nm, and (b) the optimum groove depths calculated by (9.4) and (9.5) vs. wavelength for the cases of $d = 5.0, 5.5, 6.0$ nm.

footprint of 1 m². A photomask was used, with a high groove density grating pattern of $1/D = 7600$ lines/mm and a land-to-period ratio of $a/D = 0.4$ with an exposure area of 5 mm × 5 mm (see Figure 9.1). Grating substrates were 300 μm thick, standard silicon wafers.

A unique, nonadiabatic photochemical reaction [25, 26] between the local optical near-field generated on the photomask and the photoresist was used to suppress contributions from propagating light leaking through the photomask [27, 28]. This process realized very high resolution, even when using commercial photoresist and a conventional visible-light source for fabrication. In our case, the use was made of a photoresist of FH-SP-3CL (Fujifilm Electronic Materials Co., Ltd.) and light soure of a 500-nm wavelength visible-light source.

The relief pattern developed was then used as a mask to rule the laminar-type grooves on the substrate, and was performed using an inductively coupled plasma etching method with Cl gas. Groove depths, h, were intended to be 3 nm. Next, at the Japan Atomic Energy Agency, a Mo/SiO_2 multilayer was deposited on the master gratings by use of an ion-beam sputtering method. The periodic length, d, and the ratio of (Mo thickness)/(periodic length), Γ, were set to be 5.5 nm and 0.4. Also, the number of periods, N, was 30.

To examine the period length of the multilayer, small-angle X-ray scattering measurements were performed using an X-ray diffractometer with a $CuK\alpha$ (8.05 keV) discharge source. With a standard θ–2θ scan mode measurement, the groove direction was set perpendicular to the rotation axis, to avoid diffraction peaks caused by grating grooves. The measured multilayer period length, d, was 5.33 nm.

The surface of the fabricated multilayer grating was observed using an atomic force microscope (Dimension-3000 SPM, Digital Instruments-Veeco, tip size: 15 nm diameter). Figure 9.4 shows a bird's-eye view of the grating. This observation showed us that the groove depth and rms surface roughness of the lands were 3.1 ± 0.5 nm and 1.2 ± 0.2 nm.

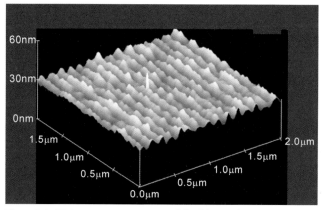

Figure 9.4 AFM image (2 μm × 2 μm) of the fabricated grating.

9.5
Simulation of Diffraction Efficiency

To simulate the diffraction efficiency of the diffraction gratings, various numerical calculation methods based on Maxwell's equations have been suggested [5, 6]. To obtain sufficient calculation accuracy it is sometimes necessary to split the grating into numerous segments, or deal with higher-order Fourier coefficients, resulting in the need to solve a huge matrix equation whose size is practically limited by the computer resources available. Furthermore, the optimal numeric calculation method that provides an accurate result with the least amount of calculation depends on the structure of the grating, e.g. blazed or laminar-type, and also on the ratio of the grating period, or depth of grooves, to the wavelength. Here, we have investigated the performance of a variety of calculation methods by comparing them with our experimental results, and also performed cross-checking of the results of each calculation method. As a result, we concluded that the rigorous coupled wave analysis (RCWA) method [29, 30] is one of the most efficient methods for analyzing laminar-type gratings used in the soft X-ray region [34].

With the RCWA method, the relevant region is divided into three sections, the incidence medium, exit medium and grating region (see Figure 9.5). Additionally, the grating region is divided into layers, which we call 'virtual mediums'. In each virtual medium, as well as in the homogeneous mediums (for example, the incidence or exit mediums), solutions for Maxwell equations are described using eigenmodes. The electromagnetic field in these virtual mediums is expressed as the superposition of the eigenmodes, and in the incidence and exit mediums is expressed as the superposition of the diffraction modes. The electromagnetic field in each medium is determined by matching the expansion coefficients with the boundary conditions between the adjustment mediums. Also, these virtual mediums consist of rectangular layers. This is why the RCWA is an efficient method of treating laminar-type

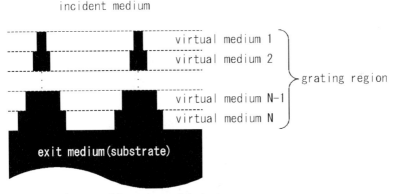

Figure 9.5 Schematic of the structure of the laminar-type multilayer grating used in rigorous coupled wave analysis (RCWA).

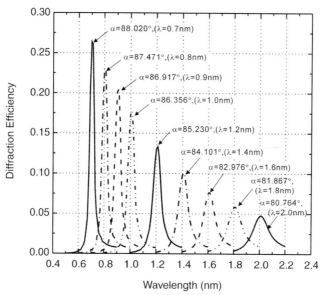

Figure 9.6 Calculated diffraction efficiency curves for the Mo/SiO$_2$ multilayer laminar-type grating of $D = 1/7600$ mm, $h = 3.1$ nm, $a/D = 0.4$, $N = 30$, $d = 5.33$ nm, $\Gamma = 0.4$, and $m_G = m_M = +1$. Incidence angles, α, as well as the wavelengths assumed for deriving them by solving (9.1) and (9.2), are shown at the respective curves.

gratings, whose structure is well approximated with rectangular layers. Another reason is that the real part of each layer's refractive index is effectively unity, making the higher-order eigenmodes, which expand the electromagnetic field in the virtual mediums, negligible.

Figure 9.6 shows the calculated diffraction efficiency curves for the Mo/SiO$_2$ multilayer laminar-type grating of $D = 1/7600$ mm, $h = 3.1$ nm, $a/D = 0.4$, $N = 30$, $d = 5.33$ nm, $\Gamma = 0.4$, and $m_G = m_M = +1$. The incidence angles, α, as well as the wavelengths assumed for deriving them by solving (9.1) and (9.2), are indicated on the respective curves.

9.6
Measurement of Diffraction Efficiency

The diffraction efficiency of the multilayer grating was measured at the evaluation beamline for soft X-ray optical elements, Ritsumeikan University, Shiga, Japan [32, 33]. This unique monochromator with a scanning mechanism based on surface normal rotation (SNR) made it possible to cover a wavelength range from 0.7 nm to 2.0 nm. SNR monochromator design principle details are described in Ref. [34]. Diffraction efficiency was calculated by dividing two types of output: the detector output

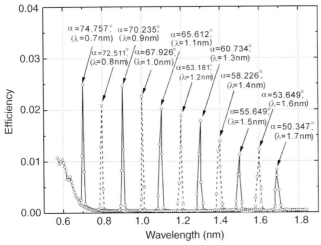

Figure 9.7 Diffraction efficiency curves of a potassium KAP crystal ($Co_2HC_6H_4CO_2K$, orientation: (0 0 1), $2d = 2.664$ nm).

produced when observing a diffracted beam from a sample grating and the detector output when observing a direct beam without the sample.

The diffraction efficiency curves of a potassium KAP crystal ($Co_2HC_6H_4CO_2K$, orientation: (0 0 1), $2d = 2.664$ nm) are shown in Figure 9.7 to illustrate the performance of the reflectometer and for the sake of comparison with the multilayer grating cases described below. Incidence angles, α, were determined by Bragg conditions for the wavelengths, λ, between 0.7 nm and 1.6 nm, assuming a refractive index of unity. The values of α and λ are shown for those curves. A peak efficiency of 0.246 was observed at 0.9 nm. The full width at half-maximum (FWHM) was from 0.010 nm to 0.016 nm. We noted that the peak efficiency and FWHM are affected by the spectral purity and vertical divergence of the incidence beam, in addition to the optical quality of the crystal itself.

Figure 9.8 shows the diffraction efficiency curves of the fabricated multilayer grating for the first spectral order, $m_G = +1$. The incidence angles, α, as well as the wavelengths assumed for deriving them by solving (9.1) and (9.2), are indicated on the respective curves. In Figure 9.8 the dashed curve in the shorter-wavelength range at the foot of the efficiency curves is the residual background due to the stray light from the monochromator. Subtracting the background intensity, a maximum diffraction efficiency of 0.032 at 0.70 nm (1.77 keV) was observed for $\alpha = 88.020°$. We also noted that diffraction efficiencies in the shorter-wavelength region are slightly higher than those of the KAP crystal (see Figure 9.7).

Peak wavelength positions on the efficiency curves coincide well with those predicted by kinematic theory and numerical simulation. However, in some curves, explicit shoulders appeared at shorter wavelengths, with an FWHM almost twice as wide as the calculated curves shown in Figure 9.6. The cause of this phenomena is not clear at this point and should be investigated in future works, including both TEM observations and subsequent simulations of diffraction efficiency based on

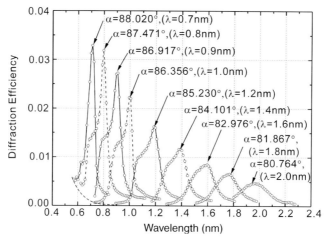

Figure 9.8 Diffraction-efficiency curves of the fabricated multilayer grating for the first spectral order, $m_G = +1$. The incidence angles, α, as well as the wavelengths assumed for deriving them by solving (9.1) and (9.2) are indicated at the respective curves.

grating and multilayer structures, which take into account the results of the TEM observations.

Figure 9.9 shows the efficiency curves of various spectral orders for the incidence angles of (a) 87.471° and (b) 85.230°, which were determined to produce peak efficiencies at 0.8 nm and 1.2 nm, for the first spectral order. The spectral order, m_G, is shown for each curve. We noted that the diffraction efficiency of zeroth and higher spectral orders is well suppressed over one order of magnitude lower than the diffraction efficiency of the first spectral order. This feature was observed for a Mo/SiO_2 multilayer-coated laminar-type grating having a groove density of

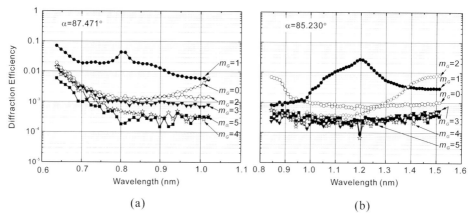

Figure 9.9 Diffraction-efficiency curves of various spectral orders, m, for the incidence angles of (a) 87.471° and (b) 85.230°.

2400 lines/mm [15], and that was made using conventional holographic and reactive-ion-etching methods. After comparing both results, we concluded that the newly fabricated multilayer grating with a groove density of 7600 lines/mm was capable of suppressing the same amount of unnecessary spectral order light as a grating with 3 times lower groove density.

9.7
Roughness Evaluation using Debye–Waller Factors

Measured diffraction efficiency is usually lower than the efficiency calculated assuming an ideal grating and multilayer structures. This is due to surface roughness at the top of and at the boundaries inside the multilayers, as well as other imperfections, including possible exotic layers generated by interdiffusion at the boundaries [35]. In this section we describe an estimation of these effects, by comparing measured diffraction efficiency with theoretical efficiency, including for simplicity, Debye–Waller factors that represent all possible imperfections as a composite roughness.

Figure 9.10 shows the comparisons of the measured and calculated diffraction efficiencies of the first spectral order. The closed circles show measured diffraction efficiencies, and the open symbols and dashed curves indicate the diffraction efficiencies calculated by a code based on the RCWA method [31] and Debye–Waller factors [36].

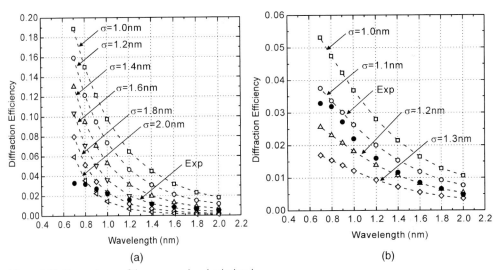

Figure 9.10 Comparisons of the measured and calculated diffraction efficiencies of the first order's light. Closed circles indicate measured diffraction efficiencies, and the open symbols and dashed curves indicate the diffraction efficiencies calculated using a code that is based on the RCWA method [31] and Debye–Waller factors [36], and is denoted by (a) (9.6) and (b) (9.7).

In Figure 9.10(a), conventional Debye–Waller factor notation was used:

$$R = R_0 \exp\left[-(4\pi\sigma\cos\alpha/\lambda)^2\right] \quad (9.6)$$

where R_0 and σ are diffraction efficiency, assuming an ideal structure and rms roughness, and that all possible imperfections are represented by σ as a composite roughness. In this case, the theoretical diffraction efficiency curves do not demonstrate a meaningful fitting with the measured diffraction-efficiency curves.

Figure 9.10(b) shows the results obtained using an improved expression of the Debye–Waller factor that considers exact momentum transfer from the boundary to the photons of the first spectral order's light:

$$R = R_0 \exp\left\{-[2\pi\sigma(\cos\alpha + \cos\beta)/\lambda]^2\right\} \quad (9.7)$$

We easily found that the theoretical diffraction efficiency curves calculated using (9.7) demonstrate a better fit to the measured diffraction efficiencies. A comparison of these measured and theoretical efficiencies suggests that the multilayer grating has a roughness of σ to 1.2 nm. Also, this result is consistent with the roughness obtained using the AFM measurement described in Section 9.4.

Acknowledgements

The authors wish to thank Dr. T. Kawazoe and Prof. M. Ohtsu of the University of Tokyo for fabricating our grating using optical near-field lithography, and Drs. V. Yashchuk and D. Voronov at the Optical Metrology Laboratory of the Advanced Light Source, Lawrence Berkeley National Laboratory, for their AFM measurements.

References

1 Tsuji K., Injuk J. and Grieken R.V. (2004) *X-Ray Spectrometry: Recent Technological Advances*, WILEY-VCH, Weinheim.

2 Samson J.R. and Ederer D.L. (1998) *Vacuum Ultraviolet Spectroscopy I in the Experimental Methods in the Physical Sciences*, Vol. 31 Academic Press, New York, or Samson J.R. and Ederer D.L. (1998) *Vacuum Ultraviolet Spectroscopy II in the Experimental Methods in the Physical Sciences*, Vol. 32 Academic Press, New York.

3 Kinoshita, T., Tanaka, Y., Matsukawa, T., Aritani, H., Matsuo, S., Yamamoto, T., Takahashi, M., Yoshida, H., Yoshida, T., Ufuktepe, Y., Nath, K.G., Kimura, S. and Kitajima, Y. (1998) *Journal of Synchrotron Radiation*, **5**, 726–728.

4 Ishiguro, E., Maezawa, H., Sakurai, M., Yanagihara, M., Watanabe, M., Koeda, M., Nagano, T., Sano, K., Akune, Y. and Tanino, K. (1992) *Proceedings of SPIE*, **1739**, 592–603.

5 Lowen, E.G. and Popov, E. (1997) *Diffraction Gratings and Applications in Optical Engineering*, Vol. 58 (ed. B.J. Thompson), Marcel Dekker, New York.

6 Neviere, M. and Popov, E. (2002) *Light Propagation in Periodic Media: Differential Theory and Design in Optical Engineering*, Vol. 81 (ed. B.J. Thompson), Marcel Dekker, New York.

7 Underwood, J.H., Gullikson, E.M., Koike, M. and Batson, P.C. (1997) *Proceedings of SPIE*, **3113**, 311–221.

8 Koike, M., Sano, K., Gullikson, E., Harada, Y. and Kumata, H. (2003) *Review of Scientific Instruments*, **74**, 1156–1158.

9 Heimann, P.A., Koike, M. and Padmore, H.A. (2005) *Review of Scientific Instruments*, **76**, 063102.

10 Cocco, D., Bianco, A., Kaulich, B., Schaefers, F., Mertin, M., Reichardt, G., Nelles, B. and Heidemann, K.F., (2007) *American Institute of Physics-Conference Paper*, **CP-879**, 497–500 (in CD).

11 Senf, F., Reichardt, G., Nelles, B. and Heidemann, K.F., (2007) *American Institute of Physics-Conference Paper*, **CP-879**, 918–921.

12 Ishiguro, E., Kawashima, T., Yamashita, K., Kunieda, H., Yamazaki, T., Sato, K., Koeda, M., Nagano, T. and Sano, K. (1995) *Review of Scientific Instruments*, **66**, 2112–2115.

13 Polack, F., Lagarde, B., Idir, M., Cloup, A.L., Jourdain, E., Roulliay, M., Delmotte, F., Gautier, J. and Ravet-Krill, M.-F., (2007) *American Institute of Physics-Conference Paper*, **879**, 489–492 (in CD).

14 Ishino, M., Heimann, P.A., Sasai, H., Hatakeyama, M., Takenaka, H., Sano, K., Gullikson, E.M. and Koike, M. (2006) *Applied Optics*, **45**, 6741–6745.

15 Imazono, T., Ishino, M., Koike, M., Sasai, H. and Sano, K. (2007) *Applied Optics*, **46**, 7054–7060.

16 Inao, Y., Nakazato, S., Kuroda, R. and Ohtsu, M. (2007) *Microelectronic Eng*, **84**, 705–710.

17 Ohtsu M. (2005) *Progress in Nano-electro-optics IV, Characterization of Nano-optical Materials and Optical Near-field Interactions*, Springer-Verlag, Berlin.

18 Ohtsu M. (2006) *Progress in Nano-electro-optics V, Nanophotonic Fabrications, Devices, Systems, and Their Theoretical Bases*, Springer-Verlag, Berlin.

19 Warburton, W.K. (1990) *Nucl Instr Meth*, **A291**, 278–285.

20 Henke, B.L., Gullikson, E.M. and Davis, J.C. (1993) *At Data Nucl Data Tables*, **54**, 181–342.

21 http://henke.lbl.gov/optical_constants/getdb2.html.

22 Hellwege, K.-H. (1937) *Zeitschrift Fur Physik*, **106**, 558–596.

23 Ishino, M. and Yoda, O. (2002) *Journal of Applied Physics*, **92**, 4952–4958.

24 Ito, T., Terao, A., Inao, Y., Yamaguchi, T. and Mizutani, N. (2007) *Proceedings of SPIE*, **6519**, 6519.

25 Kawazoe, T., Kobayashi, K., Takubo, S. and Ohtsu, M. (2005) *Journal of Chemical Physics*, **122**, 024715 (4 pages).

26 Kawazoe, T., Yamamoto, Y. and Ohtsu, M. (2001) *Applied Physics Letters*, **7**, 1184–1186.

27 Yonemitsu, H., Kawazoe, T., Kobayashi, K. and Ohtsu, M. (2007) *Journal of Luminescence*, **122–123**, 230–233.

28 Kawazoe, T., Ohtsu, M., Inao, Y. and Kuroda, R. (2007) *Journal of Nanophotonics*, **1**, 011595 (9 pages).

29 Moharam, M.G. and Gaylord, T.K. (1997) *Journal of the Optical Society of America*, **71**, 811–818.

30 Moharam, M.G. and Gaylord, T.K. (1983) *Journal of the Optical Society of America*, **73**, 1105–1112.

31 Miyauchi, S. and Koike, M. (2006) *Shimadzu Review*, **62**, 193–199 (in Japanese).

32 Koike, M., Sano, K., Yoda, O., Harada, Y., Ishino, M., Moriya, N., Sasai, H., Takenaka, H., Gullikson, E., Mrowka, S., Jinno, M., Ueno, Y., Underwood, J.H. and Namioka, T. (2002) *Review of Scientific Instruments*, **73**, 1541–1544.

33 Koike, M., Sano, K., Harada, Y., Yoda, O., Ishino, M., Tamura, K., Yamashita, K., Moriya, N., Asai, H., Jinno, M. and Namioka, T. (2002) *Proceedings of SPIE*, **4782**, 300–307.

34 Koike, M. and Namioka, T. (2002) *Applied Optics*, **41**, 245–257.

35 Miura, H., Ma, E. and Thompson, V. (1990) *Journal of Applied Physics*, **70**, 4287–4294.

36 Spiller, E. (1994) *Soft X-ray Optics*, SPIE Press, Bellingham.

10
Nanostructuring of Thin-Film Surfaces in Femtosecond Laser Ablation
Kenzo Miyazaki

10.1
Introduction

During the last two decades, it has been demonstrated that intense femtosecond (fs) laser pulses are extremely effective for high energy-density excitation of solid surfaces and resulting precision processing of materials [1, 2]. The advantage of using fs-laser pulses in material processing is the fact that the ultrafast interaction allows us to make efficient energy deposition onto the target and greatly suppress undesirable effects such as thermal and mechanical damage of the material. In such light–matter interactions, however, the spatial resolution is usually limited to the order of laser wavelength λ_0 by the diffraction limit. Much attention has been focused on optical near-fields [3, 4] and nonlinear interaction processes [1, 2] to overcome this limit and achieve high-resolution processing on a nanometer level, because of their abilities of reducing the effective interaction volume.

On the other hand, the generation of periodic nanostructures has often been observed in fs-laser ablation experiments for a variety of materials, including ceramics [5–14], insulators [15–22], semiconductors [23–31], and metals [32, 33]. In these experiments, the ablation trace of nanometer size appears to be spontaneously produced on the target surfaces. Much attention has been focused on the nanostructuring, because the results suggest potential applications of fs lasers to nanoscience and nanotechnology. The characteristic properties of nanostructuring have been investigated for a broad range of laser parameters such as fluence, polarization, wavelength, and number of pulses, as well as the physical and chemical properties of each target material. Depending on the material and experimental conditions, some of nanostructures observed may be attributed to the well-known *ripple* with a structure size of the order of λ_0, which has been considered as a universal phenomenon induced by the interference between the incident field and the scattered surface wave [34–36].

In contrast to the ripple, most nanostructures created with fs-laser pulses, especially on ceramics and insulators, represent the period d much less than λ_0, being often as small as $d \sim \lambda_0/10$ or ~ 10 nm [7, 8, 11]. The nanostructure formation is observed with

Nanophotonics and Nanofabrication. Edited by Motoichi Ohtsu
Copyright © 2009 WILEY-VCH Verlag GmbH & Co. KGaA, Weinheim
ISBN: 978-3-527-32121-6

superimposed number of fs-laser pulses at moderate fluences around the ablation threshold. The observation of nanostructuring has stimulating many studies to understand the fascinating phenomenon, since the ultrafast interaction may provide a new approach to the ultimate high-precision material processing that has been never accessible with longer laser pulses. Several authors have proposed and discussed possible interaction processes of nanostructuring on the basis of self-organization [16, 18, 21, 22, 24], change in refractive index [17], second-harmonic generation [19, 23], Coulomb explosion [9], and surface energy change [37]. However, none of them appears to be successful in providing a comprehensive and/or versatile picture of the interaction mechanism.

The physics of nanostructuring has been an interesting subject for the purposes of possible development of nanoscale processing with fs lasers. Here, we review our recent experimental studies [7, 8, 11, 38–41] of periodic nanostructure formation on hard thin films such as diamond-like carbon (DLC) and TiN, concentrating on its interaction processes.

10.2
Experimental

In the experiments, the authors employed two Ti:sapphire fs-laser systems using the chirped-pulse amplification technique, both of which were operated at a repetition rate of 10 Hz. One of them produces a peak power of 1 TW in 40-fs pulses at $\lambda_0 \sim 800$ nm. The frequency doubling and tripling of 800-nm pulses produce a peak power of 0.2 TW in 60-fs pulses at $\lambda_0 \sim 400$ nm and 20 GW in ultraviolet (UV) 140-fs pulses at $\lambda_0 \sim 267$ nm, respectively. The second fs-laser system has been developed for the study of material processing and can produce an output with a well-defined intensity distribution and good temporal characteristics. The laser system produces 100-fs pulses at 10 Hz, where the pulse duration has been designed to be relatively long for the target experiment.

Throughout the experiments, the linearly or circularly polarized fs-laser pulses were focused in air on a thin film with a 100-cm focal-length lens, where the polarization control was made with half and quarter waveplates. The laser fluence F used in the experiments was usually less than the single-pulse ablation threshold.

As the target for ablation, the authors used DLC, TiN and CrN films deposited on stainless steel plates or Si substrates. The surface morphology was observed with a scanning electron microscope (SEM) and a scanning probe microscope (SPM).

10.3
Properties of Nanostructuring

The nanostructure formation on hard thin films was observed by changing laser polarization, superimposed number of pulses N, fluence F, wavelength λ_0, and pulse width to make clear the experimental conditions for nanostructuring.

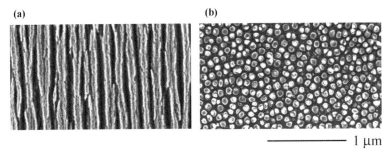

Figure 10.1 SEM images of the DLC film ablated with (a) $N = 300$ of linearly (horizontally) polarized fs-laser pulses at $F = 0.12$ J/cm^2 and (b) $N = 30$ of circularly polarized pulses at $F = 0.16$ J/cm^2.

10.3.1
Polarization

Figure 10.1 shows typical examples of the SEM image of DLC film surface ablated with superimposed shots of fs pulses with different polarizations. The linearly polarized pulses produce the ablation traces with a periodic array of line-like structures, which are highly oriented to the direction perpendicular to the laser polarization with the mean size of period $d \sim 130$ nm. In contrast, the circularly polarized pulses produce a dot-like structure with the mean grain size of $\phi \sim 100$ nm. The nanostructures have also been observed for TiN and CrN films with comparable sizes of d and ϕ [7, 11]. The observation indicates that the laser E-field direction plays an essential role in the nanostructure formation.

10.3.2
Multiple Pulses

It has been observed that the nanostructure is formed only with a superimposed number of fs-laser pulses, and a single laser pulse is never able to produce such a fine structure. The number of pulses N required for nanostructuring depends on the fluence F, which is usually less than a single pulse ablation threshold. The effect of N on the nanostructure formation has been studied in detail for DLC and TiN, and the results are shown and discussed in each subsection.

10.3.3
Fluence

Figure 10.2 shows the SEM images of TiN film ablated with multiple shots of linearly polarized fs pulses at different fluences. The period size d is observed to increase with increasing F. The result was used to deduce the ablation threshold F_{th} [11]. The estimated F_{th} of TiN is 0.14 J/cm^2 for $N = 100$ of 40-fs pulses, being in excellent agreement with that in the previous measurement [6]. When F is increased from F_{th}, d increases rapidly and tends to saturate at the ripple size of $d \sim \lambda_0$. The results for TiN

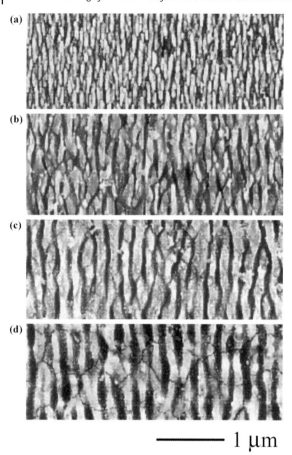

Figure 10.2 SEM images of the TiN film ablated with multiple shots of linearly polarized 40-fs-laser pulses at (a) $F = 0.15$ J/cm^2, (b) 0.18 J/cm^2, (c) 0.23 J/cm^2, and (d) 0.98 J/cm^2. The polarization is horizontal.

have shown that d or ϕ of less than $\sim\lambda_0/5$ is produced only in a region of $F/F_{th} < \sim 1.2$. The strong F-dependence of nanostructuring was also observed for CrN and DLC films. This F-dependence was almost independent of laser polarizations.

The fluence F for nanostructuring also depends on each material. For example, F to produce $d < \sim\lambda_0/2$ is $F = 0.14 \sim 0.23$ J/cm^2 for TiN, $F = 0.11 \sim 0.16$ J/cm^2 for DLC, and $F = 0.11 \sim 0.12$ J/cm^2 for CrN at $N = 100$ of 40-fs pulses [8, 11]. The very narrow region of F for CrN is due most likely to its low melting temperature, in comparison with those of TiN and DLC. In addition, the thermal stability of CrN would be much lower than that of TiN. Thus, CrN is more sensitive to F than TiN, resulting in its low value of $F_{th} \sim 0.10$ J/cm^2.

These results suggest that the nanostructuring is strongly affected by the intrinsic thermal properties of material, and the nonthermal process should be essential [38].

10.3.4
Laser Wavelength

The ablation experiment for TiN and DLC was made with the UV fs-laser pulses at 267 nm to see the effect of λ_0 on the nanostructure formation [7]. The results have shown that the fine periodic structures consisting of the line-like and dot-like elements can also be produced with multiple shots of linearly and circularly polarized UV pulses at $F \sim F_{th}$, respectively. The effect of laser polarization on the surface structure was the same as that for the 800-nm pulses. However, the structure sizes were reduced to $d \sim 40$ nm ($\phi \sim 45$ nm) for TiN and $d \sim 30$ nm ($\phi \sim 25$ nm) for DLC. It is noted that the size d and ϕ of nanostructures is almost proportional to λ_0 used for ablation [7].

10.3.5
Pulse Width

The nanostructure formation has been observed so far only with ultrashort laser pulses on a fs time scale, and ns and longer laser pulses can never induce it. This was confirmed with the ablation experiment for TiN and DLC [7, 42], where the thin-film targets were irradiated at a low fluence of 20-ns, 532-nm pulses from a Q-switched Nd:YAG laser, using the same focusing optics as those for the fs-laser ablation. The results have shown that the 20-ns-laser pulses never form nanostructures on these film surfaces. The ns-laser pulses induced cracks and pits on the ablated TiN surface, and the DLC surfaces were partially melted and spalled, due most likely to the repetitive thermal shocks, in the ablated area.

10.4
Bonding-Structure Change

In the ablation experiment with fs-laser pulses, the authors have found that DLC surface is selectively modified into a glassy carbon (GC) layer under almost the same experimental conditions as for nanostructuring [8, 38]. DLC is an amorphous material consisting of a mixture of tetrahedral (sp^3) diamond structure and trigonal (sp^2) graphite structure, whereas GC is a kind of graphite having a small crystallite size of about 3 nm with the sp^2 structure [43].

Figure 10.3 represents the Raman spectra of DLC films irradiated with 800-nm and 267-nm fs pulses, where the spectrum of nonirradiated DLC film is shown for comparison. The Raman spectra of carbonaceous materials are well known, as given in the literature [43, 44]. For the argon-ion laser line at 514.5 nm, the DLC spectrum consists of an asymmetric shape with a broad peak around 1530 cm^{-1}, as seen in Figure 10.3(a), due to a mixture of the sp^3 and sp^2 bonding structures, while the sp^3 diamond structure has a sharp spectral peak at 1333 cm^{-1}. On the other hand, the Raman spectrum of GC includes two spectral peaks at 1355 and 1590 cm^{-1} where the former peak is much larger than the latter, as shown in Figure 10.3(b) and (c).

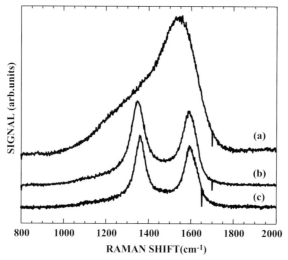

Figure 10.3 Raman spectra of DLC films: (a) nonirradiated, (b) irradiated with $N=100$ of 40-fs, 800-nm laser pulses at $F=0.13$ J/cm^2, and (c) irradiated with $N=100$ of 140 fs, 267-nm laser pulses at $F=0.12$ J/cm^2.

The initial DLC spectrum was observed to slowly move to the GC with an increase in F from $F_{th} \sim 0.11$ J/cm^2. This change in the spectrum would be due to the selective transition of sp^3 in DLC to sp^2. At a higher fluence of more than $F \sim 0.16$ J/cm^2, two peaks in the GC spectrum was observed to become shallow again, suggesting a remixture of sp^3 and sp^2 structures due likely to recrystallization [38].

The GC is usually manufactured as a bulk material by means of carefully controlled heating of polymeric resin [45], and no technology has been developed so far to grow its thin film. The experiment demonstrates that fs-laser pulses can be used for *direct and selective* modification from DLC to GC via the dissociative transition of sp^3 to sp^2, suggesting a promising approach to a thin-film technology of GC. It has been observed that the bonding-structure change is independent of polarization and wavelength of fs-laser pulses.

10.5
Dynamic Processes

In the fs-laser ablation, the incident laser pulse energy usually localizes in the target material to induce direct solid–vapor or –plasma transition followed by a rapid volume expansion of material [1]. For example, the fs-laser pulse incident on an opaque dielectric material is absorbed by bound electrons in the surface layer, followed by the ionization of material and the subsequent acceleration of free electrons through the inverse Bremstrahlung. The accelerated free electrons interact with the electron subsystem and induce avalanche ionization to rapidly increase the free-electron density N_e in the surface layer. After the end of interaction, the free electrons are quickly cooled down through the energy transfer to the lattice. The time scale of this electron relaxation is known to be of the order of picosecond [1]. The energy transfer to the lattice should be accompanied with a bonding-structure change and/or the bond breaking for melting and ablation.

10.5.1
Reflectivity of Ablating Surface

In the ultrafast interaction on the thin films concerned, free electrons created with fs-laser pulses should play the principal role in nanostructuring. The increase in N_e leads to a rapid change in the refractive index n of target materials. The refractive index change can be observed through a change in the reflectivity R of material surface [46]. Using the simple Drude model, the refractive index n including N_e may be described as $n = n_0(1-\omega_p^2/\omega^2)^{1/2}$, where n_0 is the refractive index of a material, $\omega_p = [e^2 N_e/(\epsilon_0 m)]^{1/2}$ is the plasma frequency with charge e and mass m of an electron, and ω is the laser frequency. When the laser pulse incident on a target with $n>1$ is so intense to produce N_e larger than the critical density at $\omega_p = \omega$, n of the surface layer is decreased rapidly. With the decrease in n, $R = [(1-n)/(1+n)]^2$ decreases first to $R \sim 0$ and increases up to $R \sim 1$. The characteristic change in R was used to study the dynamic process on the thin-film surface [38].

DLC film deposited on stainless steel plate was used for the measurement of R with 100-fs, 800-nm laser pulses. The reflectivity R for a single laser pulse is the ratio of the reflected pulse energy to the incident energy. This ratio was averaged for the superimposed multiple pulses and measured as functions of F and N. The results were compared with the surface morphological change following irradiation of the fs-laser pulses, as well as the bonding-structure change observed with the Raman spectroscopy.

Figure 10.4 shows R measured as a function of F for $N=1$ and $N=100$. At the nonirradiated DLC surface, R is about 14%. For a single pulse, R was observed to increase up to $R \sim 0.2$ at $F = 0.46 \, \text{J/cm}^2$. This increase in R is ascribed to the generation of free electrons to induce $n < 1$ in the DLC, since the ablation itself produces no effect on R in the single-pulse experiment. With increasing F, the critical density N_c for $\omega_p \sim \omega$ is created more rapidly in the temporal evolution of a laser pulse, leading to an increase in the pulse energy to be reflected.

On the other hand, R for $N=100$ in Figure 10.4 slightly increases with increasing F and starts to decrease at $F \sim 0.12 \, \text{J/cm}^2$. This decrease in R shows that a certain accumulation of laser pulse energy or *incubation effect* takes place in the target, despite the low repetition rate of 10 Hz. The incubation effect is a phenomenon that a part of the incident pulse energy is effectively stored shot by shot in the target, which is known to reduce F_{th} with an increase in N [47, 48]. Then, even if F is smaller than F_{th} for the single-pulse ablation, the multiple shots of laser pulses can initiate the ablation.

Figure 10.5 shows the images of the central target area of $1 \, \mu m \times 1 \, \mu m$ irradiated with $N=100$, where (a)–(d) correspond to the targets indicated as (A)–(D) in Figure 10.4, respectively. The image (a) of surface (A) represents no appreciable

10 Nanostructuring of Thin-Film Surfaces in Femtosecond Laser Ablation

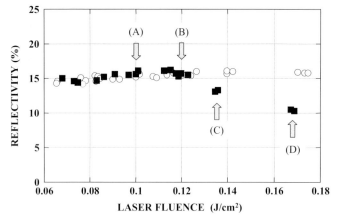

Figure 10.4 Reflectivity measured as a function of fluence for a single pulse, $N = 1$ (open circles) and for the multiple pulses, $N = 100$ (solid rectangles). The SPM images and the Raman spectra shown in Figures 10.5 and 10.6 are observed for the targets (A)–(D) for $N = 100$.

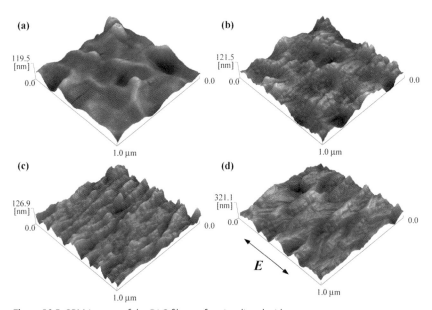

Figure 10.5 SPM images of the DLC film surface irradiated with $N = 100$ of fs-laser pulses at (a) $F = 0.10 \, \text{J/cm}^2$, (b) $0.12 \, \text{J/cm}^2$, (c) $0.14 \, \text{J/cm}^2$, and (d) $0.17 \, \text{J/cm}^2$. The scan areas of images (a)–(d) are $1 \, \mu\text{m} \times 1 \, \mu\text{m}$ in the central part of the ablated targets (A)–(D) in Figure 10.4, respectively.

change in the surface morphology, while the original surface contains some roughness. The periodic ablation traces start to be observed on surface (B), where the size of period d is much smaller than λ_0. The nanostructure of $d \sim 100$ nm is clearly seen in the image (c). Since the fine structure less than λ_0 is not able to induce any pronounced change in R, the small decrease in R for surface (C) in Figure 10.4 is due most likely to an increase in the energy deposition for nanostructuring. With a further increase in F, the structure size on the surface becomes larger, as seen in Figure 10.5(d).

As mentioned above, the fs-laser pulses induce a change in the bonding structure of DLC under the same conditions as for the nanostructure formation. To see the correlation between the bonding-structure change and the nanostructuring, the Raman spectra were observed for the targets used in the measurement of R. The results are shown in Figure 10.6, where each spectrum (a)–(d) corresponds, respectively, to the target (A)–(D) in Figure 10.4 and (a)–(d) in Figure 10.5. The Raman spectrum in Figure 10.6(a) for the target (A) shows that the DLC has undergone a certain structural change, while no appreciable change in the surface morphology is observed in Figure 10.5(a). This indicates that, even at $F < F_{th}$, the N_e produced is so high as to induce the transition from DLC to GC. As seen in Figures 10.6(b), (c) and (d), an increase in F increases the transition from sp^3 bonds in DLC to sp^2 in GC. Since the dissociation energy for the C–C bond of sp^3 is smaller than that of sp^2, the structural change appears to take place in advance of the nanostructuring through the *nonthermal* process [38].

At a higher fluence of $F > \sim 0.2$ J/cm^2, the Raman spectrum was observed to reduce two peaks and have a shallower valley between those, which was similar to the shape

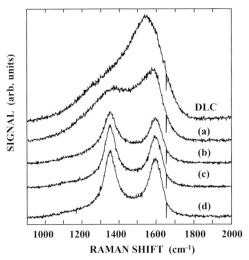

Figure 10.6 Raman spectra of the DLC surface irradiated with $N = 100$ of fs-laser pulses at (a) $F = 0.10$ J/cm^2, (b) 0.12 J/cm^2, (c) 0.14 J/cm^2, and (d) 0.17 J/cm^2, where the spectrum of nonirradiated DLC is shown for reference. The spectra (a)–(d) are for the targets (A)–(D) in Figure 10.4, respectively.

shown in Figure 10.6(a). This suggests that a high fluence leads to a mixed structure of sp^2 and sp^3 bonds, due most likely to a thermal process for recrystallization [49].

Thus, the measurements of R demonstrate that the surface nanostructuring is closely associated with the coherent production of free electrons in the surface layer and resulting ultrafast transition of the bonding structure [38].

10.5.2
Ultrafast Dynamics

The bonding-structure change should be a clue to understand the ultrafast interaction process responsible for nanostructuring. To see the role of bonding-structure change in nanostructuring, the measurement of R was made with the pump and probe technique for DLC [39]. The target used in this experiment was a 1.6-μm thick DLC film deposited on Si substrate, the surface roughness of which was $\sim$15 nm. The output from the 100-fs, 800-nm Ti:sapphire laser was split into two beams to produce a pump and a probe pulse. The pump pulse was focused for ablation of the target at normal incidence, while the probe pulse was incident at the angle $\theta = 4°$ to measure R. The pump fluence F used was less than the single-pulse ablation threshold of DLC, while the probe fluence was less than 1/300 of the pump so that the probe never induced any structural change in DLC. The time delay Δt between the pump and probe pulses was changed in a range of $\Delta t = 0$–70 ps.

Figure 10.7(a) shows R measured with the probe pulse as a function of N at $\Delta t = 0$ for parallel and crossed polarizations of the pump and probe, and at $\Delta t = 0.2$ ps for parallel polarizations. It is noted that R at $\Delta t = 0$ for parallel polarizations shows two peaks with an increase in N. For higher F, these peaks of R were observed to shift to smaller values of N. Such enhancement of R is never observed with the crossed polarizations or at $\Delta t \geq 0.2$ ps even for parallel polarizations. The result suggests that the enhancement of R is induced by a coherent interaction of the probe pulse with the film surface excited by the pump pulse. In addition, it is clear that the enhancement of R is closely associated with the incubation effect.

To see the morphological change of surfaces as a function of N, the SPM images were observed for the targets irradiated at the same F as in Figure 10.7(a). The results are plotted in Figure 10.7(b) for the mean spacing or period size d and the depth h in the nanostructure at $\Delta t = 0$ for parallel polarizations under the same conditions as in (a), and the high- and low-resolution SPM images of the targets are shown for (a) $N = 0$, (b) $N = 130$, and (c) $N = 500$ in Figure 10.8. The upper image in Figure 10.8 (b) for $N = 130$ represents a grating pattern, while no ablation is induced to produce the nanostructure on the surface, as seen in the lower image. On scanning the probe microscope over a broad area, the grating structure was found to result from the *swelling* of the DLC surface. The spacing between the fringes or crests is measured to be $d_L \sim 11$ μm with a height of $\sim$100 nm, which corresponds to $d_L = \lambda_0/\sin\theta$ for $\lambda_0 = 800$ nm and the angle $\theta = 4°$ between the pump and the probe beams. This swelling is certainly due to the bonding-structure change from DLC to GC, which decreases the mass density [50, 51]. With an increase in N, the grating groove is observed to become shallow due to the ablation initiated at the crest of fringes. The

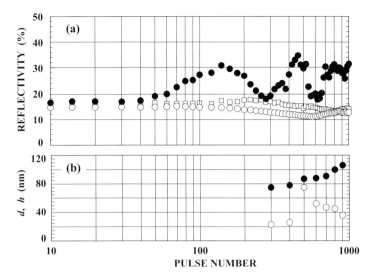

Figure 10.7 (a) Reflectivity measured as a function of pulse number N at the time delay $\Delta t = 0$ for parallel (solid circles) and perpendicular (open circles) polarizations, and at $\Delta t = 0.2$ ps (open squares) for parallel polarizations, where the pump and probe fluences are 0.14 J/cm^2 and 0.2 mJ/cm^2, respectively. (b) Mean spacing d (solid circles) and depth h (open circles) of periodic nanostructures as a function of N at $\Delta t = 0$ for parallel polarizations under the same conditions as in (a).

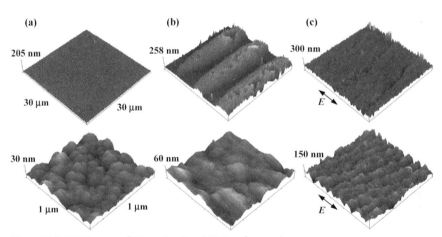

Figure 10.8 SPM images of (a) nonirradiated DLC surface, and those irradiated with (b) $N = 130$ and (c) $N = 500$ of fs-laser pulses at $\Delta t = 0$ for parallel polarizations, where the pump and probe fluences are 0.14 J/cm^2 and 0.2 mJ/cm^2, respectively, and the polarization direction is indicated with arrows in (c). The scan areas of the upper and lower images are $30 \,\mu\text{m} \times 30 \,\mu\text{m}$ and $1 \,\mu\text{m} \times 1 \,\mu\text{m}$, respectively.

nanostructure starts to be formed on the crest at $N \sim 300$, as shown in Figure 10.7(b), and no nanostructure is observed for $N < \sim 300$.

When the grating is formed on the film surface at $N \sim 100$, the pump pulse is partially diffracted into the direction of the reflected probe beam to increase R. The estimated increase in R due to the diffraction was in good agreement with the observed enhancement of $R \sim 33\%$ in Figure 10.7(a) [39]. The diffracted pump wave was also detected in the backward direction. These results demonstrate that the enhanced R is certainly due to the grating formation on the DLC surface.

It is noted that the grating pattern is formed with the addition of very weak probe pulses of which fluence is $\sim 1/700$ of the pump. This indicates that a small field modulation created by the probe pulse can produce a large morphological change or the grating, due most likely to the sharp threshold of F to induce the bonding-structure change from DLC to GC.

These results shown in Figures 10.7 and 10.8 suggest that the nanostructure is preceded by the bonding-structure change. To confirm this, the Raman spectra were measured for the DLC targets. The results have shown that the bonding-structure change starts to be observed at $N \sim 100$ in Figure 10.7. It should be noted that neither ablation nor nanostructure formation is induced around the first peak of R at $N \sim 130$ or less. The most distinct GC spectrum was observed at $N \sim 500$ with the second peak of R.

These results demonstrate that the bonding-structure change from DLC to GC is first induced with increasing N. This modification is observed as the surface swelling to form the grating at $N \sim 130$. With a further increase in N, the visible nanostructure starts to be formed at $N \sim 300$. The decrease in R after the first peak for $\sim 130 < N < 300$ is due to the local ablation of the swelled surface, which decreases the diffraction efficiency. This is consistent with the fact that the sp^2 bond can preferentially be ionized with fs-laser pulses due to its ionization energy being lower than that of the sp^3 [52].

Thus, it is concluded that the bonding-structure change from DLC to GC certainly precedes the ablation for nanostructuring. An additional important finding should be the fact that the weak probe field is able to initiate the bonding-structure change followed by a large morphological change and subsequent nanostructure formation on the swelled surface. This suggests that the nanostructure can be formed on the DLC surface when a local field is spontaneously generated, for example, due to the nanoscale swelling or roughness. The bonding-structure change increases the local efficiency of free-electron generation, where N_e would not be uniform in the surface layer. The coherent oscillation of electrons can produce local fields due to the nonuniform distribution of N_e in the surface [53], leading to the initiation of nanoscale ablation.

10.6
Local Fields

The pump-probe experiment has demonstrated that the nanostructuring of DLC is certainly preceded by the change in bonding structure from DLC to GC, and

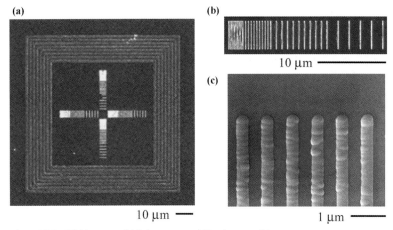

Figure 10.9 SEM images of (a) the patterned Si substrate, (b) an array of stripes, and (c) the DLC-coated stripes.

suggested that the periodic nanostructure is formed on the swelled GC surface through the nanoscale ablation induced by the local field [39]. An experiment was designed to see the generation of local fields and its role in nanostructuring of the DLC film surface [40].

For the experiment, Si substrates were prepared on which Si stripes were fabricated with electron-beam lithography and the liftoff process. Figure 10.9 shows the SEM images of the patterned Si substrate and the DLC-coated stripes. The width, length, and height of a single Si stripe are 0.1 μm, 4 μm, and 50 nm, respectively, and an array consists of the 36 parallel stripes drawn at four different periods of 200, 400, 800, and 1600 nm. The four arrays in the central area of $50 \times 50\,\mu m^2$ are set in a cross shape. The eight square-box-shape stripes surrounding the crossed arrays work as a grating for the visible light. The DLC of 900 nm in thickness was deposited on the patterned Si substrate, and then the DLC-coated stripe was ∼500 nm in width and ∼100 nm in height. The linearly polarized 100-fs, 800-nm pulses were focused at normal incidence to a spot size of 100 μm, so that the focused beam could cover the whole patterned target area of $50 \times 50\,\mu m^2$. The superimposed pulses of N 10–1000 were used at $F = 60$–$150\,mJ/cm^2$.

Figure 10.10 shows the SPM images of a patterned DLC surface irradiated with different N at $F = 70\,mJ/cm^2$, where the image of a nonirradiated stripe surface is also shown for comparison. The image in Figure 10.10(b) represents narrow line-like ablation traces formed in a small area along the crest of stripes. The period size in the nanostructure is $d \sim 60\,nm$. The line-like traces produced are in the direction perpendicular to the laser polarization. For $N = 200$, the small structure grows up on the stripe ridge, while no structure is observed on the flat surface area outside the stripe arrays. For $N = 300$, as seen in Figure 10.10(d), the distinct line-like structure with $d \sim 110\,nm$ is developed on the stripe, and the small ablation traces are also formed on the nonpatterned surface area. With a further increase in N, the ablation traces on the stripe ridge become deeper with $d \sim 180\,nm$, while the smaller structure

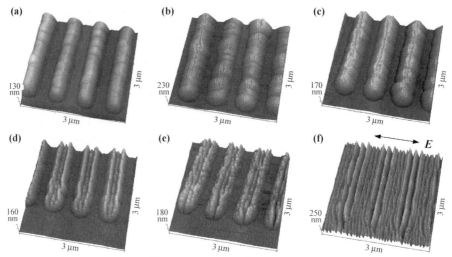

Figure 10.10 SPM images of the patterned DLC surface irradiated with (a) $N = 0$, (b) $N = 100$, (c) $N = 200$, (d) $N = 300$, (e) $N = 500$, and (f) $N = 1000$ of fs-laser pulses at $F = 70$ mJ/cm^2. The arrow indicates the polarization direction of linearly polarized pulses.

with $d \sim 90$ nm is formed on the nonpatterned area. These results indicate that the E-field intensity is locally enhanced to efficiently induce the nanoscale ablation on the stripe ridge, while F is kept below the single-pulse ablation threshold. Such enhancement of the E-field intensity preferentially takes place due to the high surface curvature along the laser polarization direction. A small enhancement would be enough to initiate the local ablation, as discussed above for the pump-probe experiment.

The laser E-field incident on the target is always parallel and perpendicular to the stripes, since the four stripe arrays are arranged in a crossed shape. Then, the effect of the different polarizations on the morphological change of the stripe surfaces could be observed under the same condition of F and N. Figure 10.11 shows the SEM images of a pair of stripes on a single target irradiated with the linearly polarized laser pulses. The line-like ablation traces are certainly formed in the direction perpendicular to the E-field. The deeper nanostructure is produced with $d \sim 110$ nm on the stripes perpendicular to the laser E-field, while the shallow nanostructure is formed with the smaller spacing $d \sim 80$ nm on the stripes parallel to the field. Thus, the local field intensity enhanced on the stripe surface depends on F and the surface shape. Figure 10.12 shows the result of d measured as a function of F for the stripe perpendicular to the laser E-field and for the nonpatterned flat surface. It is demonstrated that the F required for nanostructuring is greatly reduced on the stripe.

The morphological change of target surfaces would certainly be a fingerprint of the nanofield and resulting nanoscale ablation on the surface. The results obtained allow us to illustrate the dominant interaction process of nanostructuring. The volume swelling due to the transition from DLC to GC creates nanometer-size roughness even on a flat DLC surface. Free electrons can be efficiently produced in the GC layer

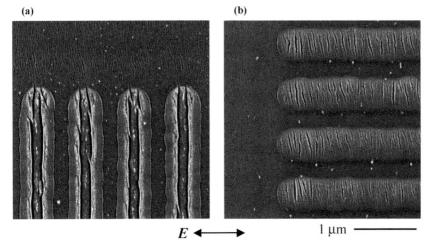

Figure 10.11 SEM images of the DLC surface including stripes (a) perpendicular and (b) parallel to the laser E-field indicated by an arrow. The laser fluence is $F = 70$ mJ/cm^2, and the number of pulses is $N = 300$.

due to the smaller bandgap than that of DLC, and N_e produced would not be uniform in the swelled GC layer on a nanometer level. In the laser field, the free electrons coherently oscillate and localize in the area much smaller than λ_0. The localized electron–hole pairs can lead to an enhancement of local field intensity along the E-field direction so as to initiate the local ablation on the surface. The enhancement of local field intensity efficiently takes place in the direction of high surface curvature along the laser E-field. This process would be similar to the enhancement of an optical near-field [3, 54].

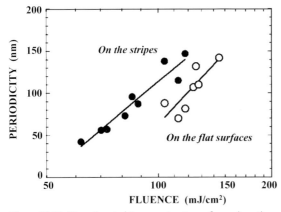

Figure 10.12 Size of period in nanostructures formed on the stripe (solid circles) and nonpatterned flat surfaces (open circles) as a function of fluence with $N = 100$ of fs-laser pulses linearly polarized along the direction perpendicular to the stripes.

10.7
Origin of Periodicity

The spontaneous generation of local fields accounts for the nanoscale ablation of DLC surface and its polarization dependence. For further understanding of nanostructuring, an important problem is the interaction to create the *nanoscale periodicity* on the surface. With the observation of the first stage of fs-laser ablation, the generation of periodicity is attributed to the excitation of surface plasmon-polaritons (SPPs) in the surface layer [41].

The target used in the experiment was a 500-nm thick DLC film deposited on polished Si substrate, the surface roughness of which was less than 1 nm. The 100-fs, 800-nm laser pulse energy was 60–170 mJ with $F = 50$–150 mJ/cm^2.

The morphological change on the flat surface was carefully observed as a function of N at a low fluence. Figure 10.13 shows the SEM images of the surface irradiated with the linearly polarized pulses of $N = 10$–1000. For $N = 10$, small ablation traces the grain size of which is less than 10 nm start to be observed, but the surface image represents neither a periodic structure nor polarization-dependent ablation trace. On increasing the shot number to $N = 30$, the randomly distributed ablation traces are unified to form line-like traces with $d = 40$–60 nm along the direction perpendicular to the laser polarization. For $N = 100$–200, the line-like traces grow in depth and length with $d \sim 120$ nm. With a further increase in N to 500–1000, the broader and deeper ablation traces are structured to show the period $d = 140$–170 nm. The similar characteristic evolution of nanostructuring was also observed for the periodic granular traces produced with the circularly polarized pulses.

The results shown in Figure 10.13 suggest the following process. In the initial stage, the bonding-structure change from DLC to GC is randomly induced in the surface layer. This structural change produces the surface roughness of nanometer size due to the random swelling of the target material. On the swelled surface with a high curvature, the local field can be generated to enhance the incident laser E-field and initiate nanoscale ablation [40]. Once the ablation is locally induced, different high curvatures are created on the surface of nanometer size. Since the field

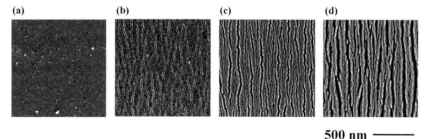

500 nm ———

Figure 10.13 SEM images of the DLC surface irradiated with (a) $N = 10$, (b) $N = 30$, (c) $N = 100$, and (d) $N = 200$ of linearly polarized fs-laser pulses at $F = 100$ mJ/cm^2. The laser polarization is horizontal.

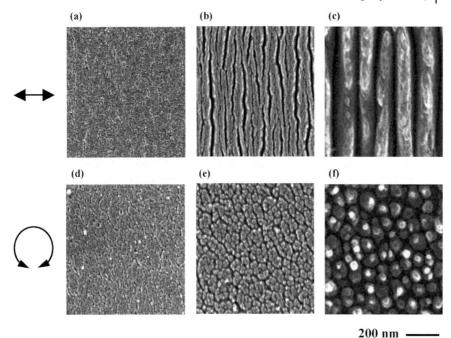

Figure 10.14 SEM images of the DLC surface irradiated with linearly (upper three) and circularly (lower three) polarized fs-laser pulses of $N = 100$ at $F = 80$ mJ/cm^2 for (a) and (d), $F = 100$ mJ/cm^2 for (b) and (e), and $F = 150$ mJ/cm^2 for (c) and (f). The laser polarization for (a)–(c) is horizontal.

enhancement should be larger in narrower spaces and valleys, the subsequent laser pulses are able to extend the length of ablation traces to combine them.

An increase in F gives almost the same effect on nanostructuring as in N [38–40]. Figure 10.14 shows the typical SEM images of an DLC surface ablated at $F = 80$–150 mJ/cm^2 of linearly and circularly polarized pulses. At the lowest fluence, the images do not represent any polarization dependence of ablation. For higher fluences, the ablation traces are structured with $d = 60$–120 nm for the linear polarization and $\phi = 40$–90 nm for the circular.

Based on the results obtained, the generation of nanoscale periodicity can be attributed to the excitation of SPPs, which results from the coherent interaction of the incident laser field with free electrons created in the surface layer. The presence of surface roughness is crucial for the SPP excitation, because on a smooth surface, the dispersion curve for the SPP lies below the photon line, and energy and momentum cannot be conserved simultaneously [55]. On the DLC surface, the bonding-structure change produces a random distribution of nanometer-size roughness in the initial stage of ablation. This surface roughness, being much smaller than λ_0, allows the incident E-field to coherently couple with the collective oscillations of free electrons to excite the SPPs in the surface layer. The SPP has a periodic change in the surface charge or resulting local field intensity to periodically initiate the nanoscale ablation of the surface.

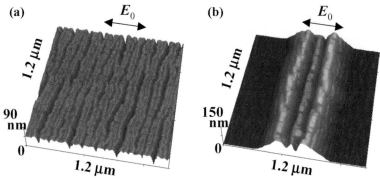

Figure 10.15 SPM images of (a) the flat surface irradiated at $F = 100$ mJ/cm^2 and for (b) the patterned surface at $F = 70$ mJ/cm^2 of linearly polarized fs-laser pulses. The superimposed number of pulses is $N = 100$ for both, and the arrows denote the polarization direction.

The plasmon wavelength λ_{sp} was estimated by referring to the results of an ablation experiment using the same patterned target as in the previous study [40]. First, the local field E_{local} spontaneously generated on the curved DLC surface was evaluated by comparing F to produce the same size of d on the curved and flat surfaces. Figure 10.15 shows the SPM images of both surfaces on which almost the same period size $d = 100$–120 nm is structured with $N = 300$ of the linearly polarized laser pulses, where $F_1 = 100$ mJ/cm^2 for the flat and $F_2 = 70$ mJ/cm^2 for the stripe. Assuming that the field intensity to produce the same size of d is the same on the two surfaces, $\Delta F = F_1 - F_2$ corresponds to the generation of an additional field $\Delta E \sim 4.0 \times 10^6$ V/cm along the laser E-field on the stripe surface. This field may be written as $\Delta E = E_{local} \cos\theta$, where E_{local} is the local field spontaneously generated by the surface charge on the stripe with the angle θ between the laser E-field and the direction perpendicular to the stripe surface [40].

The surface charge for E_{local} is coherently produced and oscillated by the incident laser field. As shown in Figure 10.12, ΔF was almost independent of F in the fluence region of interest. This suggests that the surface charge or N_e produced for initiating the ablation is almost constant for a material, while an increase in F predominantly leads to an increase in the ablation depth and area. The electron density N_e can be estimated from ΔE. Using Gauss's law [56], we have

$$\Delta E = (\varepsilon_1/\varepsilon_0 - 1) E_0 \cos^2\theta \tag{10.1}$$

where ε_0 and ε_1 are the dielectric constants of vacuum and the GC layer including the effect of free electrons, respectively. With the simple Drude model [46], ε_1 in the laser field is given by

$$\varepsilon_1 = \varepsilon_0 \left[\varepsilon_{GC} - (\omega_p^2/\omega^2)\right] \tag{10.2}$$

where ε_{GC} is the static relative dielectric constant of GC [57], and ω_p is the plasma frequency. The measured cross-section of a stripe was approximated with a semicylindrical shape, and then the angle θ was found to be 78–86° on the stripe. For this angle, (10.1) and (10.2) lead to $N_e = 0.7 - 6 \times 10^{22}$ cm^{-3}. The high electron density, in excess of the critical value $N_e = 1.7 \times 10^{21}$ cm^{-3} at $\lambda_0 \sim 800$ nm, can be generated with the evanescent field rapidly decreasing in the surface layer.

As discussed before, the free electrons are predominantly produced in the GC layer to induce a large change in ε_1. For simplicity, the film surface before ablation is assumed to consist of two layers of the upper GC and the lower DLC. When the surface is corrugated through the bonding-structure change and subsequent initial ablation, the SPPs can be excited at the interfaces between the GC and DLC and also between air and the GC. The dispersion relation $k_{sp} = k_0[\varepsilon_a\varepsilon_b/(\varepsilon_a + \varepsilon_b)]^{1/2}$ has to be satisfied for the SPP excitation, where k_{sp} and k_0 are the wave vectors of the surface plasmon and the incident light in vacuum [55], respectively, and ε_a and ε_b are the relative dielectric constants of two layers concerned. Using $\varepsilon_a = \varepsilon_1/\varepsilon_0$ for GC and $\varepsilon_b = \varepsilon_2$ for DLC [58], λ_{sp} for the GC/DLC interface was calculated as a function of N_e in the GC layer. The results are shown in Figure 10.16, where λ_{sp} for the air/GC interface is also shown for comparison. It is noted that the estimated electron density N_e leads to $\lambda_{sp} = 150$–340 nm for the GC/DLC interface, whereas λ_{sp} calculated for the air/GC interface is of the order of λ_0. The local ablation would be initiated at a period $d \sim \lambda_{sp}/2$ with the help of the local field periodically enhanced by the SPPs. The calculated period $d = 75$–170 nm for the GC/DLC interface is in good agreement with the observed size of nanostructure, where the smallest period $d \sim 75$ nm is induced with $\varepsilon_a \sim -\varepsilon_b$ at $N_e \sim 2 \times 10^{22}$ cm^{-3}.

In spite of the simple surface structure, the model based on the excitation of SPPs can account for the observed characteristic dependences of nanostructuring on laser parameters. As given by the dispersion relation, the size of periodically enhanced fields and resulting nanostructure size would be proportional to λ_0 used, being in consistent with the observations mentioned before in Ref. [7]. The SPPs can be excited for a variety of dielectric materials having nanoscale surface corrugations,

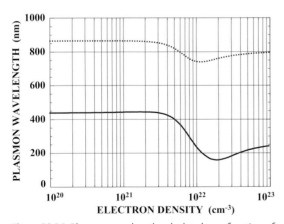

Figure 10.16 Plasmon wavelength calculated as a function of electron density for the GC/DLC interface (solid curve) and for the air/GC (dotted curve).

using fs-laser pulses, and should play a key role in the nanostructure formation on the surface.

10.8
Summary

Experimental studies have been made to understand nanostructuring of hard thin films such as DLC and TiN in the fs-laser ablation. The film surface is nanostructured at a moderate fluence of superimposed multiple fs-laser pulses, and the surface morphology strongly depends on the laser polarization. On the other hand, it has been found that the nanostructured DLC surface is modified to a GC layer due to the selective bonding-structure change induced by the fs-laser pulses.

The results of reflectivity measurements for ablating surfaces have shown the intimate relation between the nanostructuring and the bonding-structure change of DLC. The pump-probe experiment has demonstrated that the nanostructuring of DLC surface is certainly preceded by the bonding-structure change to GC and suggested that the nanoscale ablation is initiated with the local field generation on the surface.

The generation of local fields and its role in nanostructuring has been studied using the patterned DLC film. The results have demonstrated that nanoscale ablation is preferentially initiated with the enhanced local E-field on the stripe surface with high curvature.

Based on the results obtained, the origin of nanoscale periodicity has been attributed to the excitation of SPPs in the surface layer. The estimated period in nanostructures is in good agreement with the observed size, and the characteristic properties of nanostructuring observed reconcile with the excitation of SPPs. The model based on the SPPs can be applied to the nanostructuring of other materials.

Acknowledgments

The author acknowledges G. Miyaji and N. Yasumaru for their collaborations in the experimental studies, and J. Kiuchi, N. Maekawa, W. Kobayashi, and Y. Miyatani for their contributions in the experiments. This work was partially supported by the Grant-in-Aid for Scientific Research (A)18206010.

References

1 Korte, F., Adams, S., Egbert, A., Fallnich, C., Ostendorf, A., Nolte, S., Will, M., Ruske, J.-P., Chichkov, B.N. and Tünnermann, A. (2000) *Optics Express*, **7**, 41–49.

2 Ovsianikov, A. and Chichkov, B.N. (2006) *Photonics Spectra*, **40** (10), 72–80.

3 Ohtsu, M. and Kobayashi, K. (2004) *Optical Near field*, Springer, Berlin.

4 Inao, Y., Nakasato, S., Kuroda, R. and Ohtsu, M. (2007) *Microelectronic Engineering*, **84**, 705–710.
5 Bonse, J., Sturm, H., Schmidt, D. and Kautek, W. (2000) *Applied Physics A-Materials Science & Processing*, **71**, 657–665.
6 Bonse, J., Rudolph, P., Krüger, J., Baudach, S. and Kautek, W. (2000) *Applied Surface Science*, **154–155**, 659–663.
7 Yasumaru, N., Miyazaki, K. and Kiuchi, J. (2003) *Applied Physics A-Materials Science & Processing*, **76**, 983–985.
8 Yasumaru, N., Miyazaki, K. and Kiuchi, J. (2004) *Applied Physics A-Materials Science & Processing*, **79**, 425–427.
9 Dong, Y. and Molian, P. (2004) *Applied Physics Letters*, **84**, 10–12.
10 Rudolf, P. and Kautek, W. (2004) *Thin Solid Films*, **453–454**, 537–541.
11 Yasumaru, N., Miyazaki, K. and Kiuchi, J. (2005) *Applied Physics A-Materials Science & Processing*, **81**, 933–937.
12 Tomita, T., Kinoshita, K., Matsuo, S. and Hashimoto, S. (2006) *Japanese Journal of Applied Physics*, **45**, L444–L446.
13 Jia, T.Q., Zhao, F.L., Huang, M., Chen, H.X., Qiu, J.R., Li, R.X., Xu, Z.Z. and Kuroda, H. (2006) *Applied Physics Letters*, **88**, 111117.
14 Tomita, T., Kinoshita, K., Matsuo, S. and Hashimoto, S. (2007) *Applied Physics Letters*, **90**, 153115.
15 Costache, F., Henyk, M. and Reif, J. (2002) *Applied Surface Science*, **186**, 352–357.
16 Reif, J., Costache, F., Henyk, M. and Pandelov, S.V. (2002) *Applied Surface Science*, **197–198**, 891–895.
17 Wu, Q., Ma, Y., Fang, R., Liao, Y., Yu, Q., Chen, X. and Wang, K. (2003) *Applied Physics Letters*, **82**, 1703–1705.
18 Costache, F., Henyk, M. and Reif, J. (2003) *Applied Surface Science*, **208–209**, 486–491.
19 Jia, T.Q., Chen, H.X., Huang, M., Zhao, F.L., Qiu, J.R., Li, R.X., Xu, Z.Z., He, X.K., Zhang, J. and Kuroda, H. (2005) *Physical Review B-Condensed Matter*, **72**, 125429.
20 Jia, T.Q., Chen, H.X., Huang, M., Wu, X.J., Zhao, F.L., Baba, M., Suzuki, M., Kuroda, H., Qiu, J.R., Li, R.X. and Xu, Z.Z. (2006) *Applied Physics Letters*, **89**, 101116.
21 Varlamova, O., Costache, F., Reif, J. and Bestehorn, M. (2006) *Applied Surface Science*, **252**, 4702–4706.
22 Varlamova, O., Costache, F., Ratzke, M. and Reif, J. (2007) *Applied Surface Science*, **253**, 7932–7936.
23 Borowiec, A. and Haugen, H.K. (2003) *Applied Physics Letters*, **82**, 4462–4464.
24 Costache, F., Arguirova, S.K. and Reif, J. (2004) *Applied Physics A-Materials Science & Processing*, **79**, 1429–1432.
25 Daminelli, G., Krüger, J. and Kautek, W. (2004) *Thin Solid Films*, **467**, 334–341.
26 Shen, M.Y., Crouch, C.H., Carey, J.E. and Mazur, E. (2004) *Applied Physics Letters*, **85**, 5694–5696.
27 Wang, X.C., Lim, G.C., Ng, F.L., Liu, W. and Chua, S.J. (2005) *Applied Surface Science*, **252**, 1492–1497.
28 Bonse, J., Munz, M. and Sturm, H. (2005) *Journal of Applied Physics*, **97**, 013538.
29 Qian, H.X., Zhou, W., Zheng, H.Y. and Lim, G.C. (2005) *Surface Science*, **595**, 49–55.
30 Le Harzic, R., Schuck, H., Sauer, D., Anhut, T., Riemann, I. and König, K. (2005) *Optics Express*, **13**, 6651–6656.
31 Guo, X.D., Li, R.X., Hang, Y., Xu, Z.Z., Yu, B.K., Ma, H.L., Lu, B. and Sun, X.W. (2008) *Materials Letters*, **62**, 1769–1771.
32 Vorobyev, A.Y., Markin, V.S. and Guo, C. (2007) *Journal of Applied Physics*, **101**, 034903.
33 Zhao, Q.Z., Malzer, S. and Wang, L.J. (2007) *Optics Letters*, **32**, 1932–1934.
34 Guosheng, Z., Fauchet, P.M. and Siegman, A.E. (1982) *Physical Review B-Condensed Matter*, **26**, 5366–5381.
35 Sipe, J.E., Young, J.F., Preston, J.S. and van Driel, H.M. (1983) *Physical Review B-Condensed Matter*, **27**, 1141–1154.
36 Young, J.F., Preston, J.S., van Driel, H.M. and Sipe, J.E. (1983) *Physical Review B-Condensed Matter*, **27**, 1155–1172.
37 Kautek, W., Rudolph, P., Daminelli, G. and Krüger, J. (2005) *Applied Physics A-Materials Science & Processing*, **81**, 65–70.
38 Miyazaki, K., Maekawa, N., Kobayashi, W., Yasumaru, N. and Kiuchi, J. (2005) *Applied*

Physics A-Materials Science & Processing, **80**, 17–21.

39 Miyaji, G. and Miyazaki, K. (2006) *Applied Physics Letters*, **89**, 191902.

40 Miyaji, G. and Miyazaki, K. (2007) *Applied Physics Letters*, **91**, 123102.

41 Miyaji, G. and Miyazaki, K. (2008) *Optics Express*, **16**, 16265–16271.

42 Yasumaru, N., Miyazaki, K., Kiuchi, J. and Magara, H. (2003) *Proceedings of SPIE*, **4830**, 521–525.

43 Yoshikawa, M., Nagai, N., Matsuki, M., Fukuda, H., Katagiri, G., Ishida, H., Ishitani, A. and Nagai, I. (1992) *Physical Review B-Condensed Matter*, **46**, 7169–7174.

44 Nemanich, R.J. and Solin, S.A. (1979) *Physical Review B-Condensed Matter*, **20**, 392–401.

45 Dekanski, A., Stevanovic, J., Stevanovic, R., Nikolic, B.Z. and Jovanovic, V.M. (2001) *Carbon*, **39**, 1195–1205, and references therein.

46 Shank, C.V., Yen, R. and Hirlimann, C. (1983) *Physical Review Letters*, **50**, 454–457.

47 Bonse, J., Wrobel, J.M., Krüger, J. and Kautek, W. (2001) *Applied Physics A-Materials Science & Processing*, **74**, 89–94.

48 Bonse, J., Baudach, S., Krüger, J., Kautek, W. and Lenzner, M. (2002) *Applied Physics A-Materials Science & Processing*, **74**, 19–25.

49 Prawer, S., Kalish, R. and Abel, M. (1986) *Applied Physics Letters*, **48**, 1585–1587.

50 Yasumaru, N., Miyazaki, K., Kiuchi, J. and Magara, H. (2004) *Proceedings of SPIE*, **5662**, 755–759.

51 Kononenko, T.V., Kononenko, V.V., Pimenov, S.M., Zavedeev, E.V., Konov, V.I., Romano, V. and Dumitru, G. (2005) *Diamond and Related Materials*, **14**, 1368–1376.

52 Robertson, J. (1986) *Advances in Physics*, **35**, 317–374.

53 Bozhevolnyi, S.I. (2001) *Optics of Nanostructured Materials* (eds V.A. Markel and T.F. George), John Wiley & Sons, New York, Chapter 3.

54 Plech, A., Kotaidis, V., Lorenc, M. and Boneberg, J. (2006) *Nature Physics*, **2**, 44–47.

55 Raether, H. (1988) *Surface Plasmons on Smooth and Rough Surfaces and on Gratings*, Springer-Verlag, Berlin, p. 5.

56 Jackson, J.D. (1999) *Classical Electrodynamics*, John Wiley & Sons, New York, Section 1.3.

57 Williams, M.W. and Arakawa, E.T. (1972) *Journal of Applied Physics*, **43**, 3460–3463.

58 Alterovitz, S.A., Savvides, N., Smith, F.W. and Woollam, J.A. (1997) *Handbook of Optical Constants of Solids* (eds E.D. Palik and G. Ghosh), Academic, London, pp. 837–852.

11
Quantum Dot Nanophotonic Waveguides

Lih Y. Lin and Chia-Jean Wang

Although the demand for higher processor chip speeds in a computer system continues to drive the bandwidth and density requirements for on-chip and chip-to-chip interconnects, using electrons as the information carrier is now facing the challenges of speed, power consumption and crosstalk due to electromagnetic interference between channels. On the other hand, photonic integrated circuits have emerged as a promising alternative by transmitting information using photons. In general, smaller and faster photonic integrated circuit systems with versatile functionalities are desirable for applications not only in computation, but also in other areas such as optical communication and sensing.

The development of photonic integrated circuits has, nonetheless, been hindered by the diffraction limit, which restricts the extent of optical wave confinement to $\sim\lambda/2$ for guiding and transmitting optical energy in conventional optical waveguides, and constrains the integration density one can achieve using photons as information carriers. While electronic integrated circuit technologies are now firmly on the 32-nm lithographic resolution roadmap, the diffraction limit must be overcome for photonics to successfully complement and perhaps replace electrical components. Advances in nanofabrication technologies have facilitated the miniaturization and integration density of device technologies. Optical waveguides with nanoscale structures such as photonic crystals [1–3] and silicon slot waveguides [4–6] have been demonstrated to transmit optical waves with high confinement. One-dimensional nanostructures such as nanowires, nanotubes [7–9] and 0-dimensional structures such as semiconductor quantum dots (QD) and metal nanoparticles (NP) have also offered promising routes towards waveguiding on the nanoscale. Among the various technologies, transmitting optical energy by surface plasmon propagation has enabled true subdiffraction waveguiding [10–13]. Plasmonic waveguide structures that have been proposed thus far include 1D negative-index optical fibers [14], metal nanowires [15, 16], metal strips [11, 17], triangular metal wedges [18], metal slots [10, 19, 20] and metal NPs [21–26].

Nanophotonics and Nanofabrication. Edited by Motoichi Ohtsu
Copyright © 2009 WILEY-VCH Verlag GmbH & Co. KGaA, Weinheim
ISBN: 978-3-527-32121-

The chief challenge for plasmonic waveguides is the high transmission loss due to strong damping of the propagating surface plasmon. Integration of the plasmonic waveguides with active semiconductor devices such as lasers or photodetectors may also be complex. On the other hand, QDs are semiconductor materials with dimensions in nanometers that may provide a route to lower-loss devices. As in bulk semiconductors, population inversion can be achieved with a sufficient excitation rate and energy, which results in an emission coefficient greater than the absorption coefficient leading to net optical gain. Furthermore, the emission of QDs is spectrally sharp due to the 3D electron–hole pair (exciton wavefunction) confinement under excitation, which results in a sharp gain spectrum and therefore high gain and high quantum efficiency. The spectral response of QDs can easily be tuned by particle size and composition. These unique properties make QDs ideal materials for optoelectronic applications [27].

Conventional QD photonic devices utilize self-organization at the interface of strained epitaxially grown III-V or II-VI semiconductor layers to create QD nanostructures. Such a process is often not compatible with silicon CMOS fabrication. Another process to fabricate QDs is through chemical synthesis. The resulting QDs are in colloidal form and have been widely used as fluorescent tags in biomedicine [28]. In this chapter, we discuss subdiffraction-limited optical waveguides utilizing gain-enabled colloidal QDs. These QDs are conjugated with streptavidin or carboxylated for fabrication into a waveguide pattern on a silicon wafer through DNA-mediated [29] or other molecular self-assembly processes [30]. The goal is to achieve low-loss optical energy transmission with the optical gain component [30–32] and low crosstalk to enable high integration density utilizing near-field optical coupling between adjacent QDs [32, 33].

In Section 11.1, we describe the principle of the QD waveguides and modeling results. Finite-difference time-domain (FDTD) and Monte Carlo simulations are employed as modeling tools to simulate the behavior of the device. The self-assembly fabrication processes are described in Section 11.2. The advantages and drawbacks of DNA-directed and aminopropyltriethoxysilane (APTES)-mediated methods will be discussed. In Section 11.3, the current performance of the QD waveguides is reported to demonstrate the operation of the devices.

11.1
Conceptual Formation and Modeling of the Device

Figure 11.1(a) illustrates the principle of the QD waveguide. The QDs are anchored to the surface by self-assembly chemistries to form a waveguide. The pump laser, whose energy is specified to be equal or greater than the second excited state for the QD, is directed overhead to excite the quantum-confined electrons into the conduction band. By rapid relaxation, the electron–hole (e–h) pairs come to occupy the first conduction and valence band states. The input signal, with energy matched to the separation of the first electron and hole states, stimulates photon emission

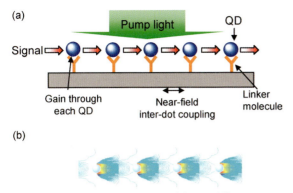

Figure 11.1 (a) Schematic drawing of the subdiffraction QD waveguide, showing its operating principle. (b) Intensity distribution of the stimulated emitted photons by FDTD simulation [30].

from the QD, which will then cascade through the waveguide. FDTD simulation of the intensity distribution of the stimulated emitted photons (Figure 11.1(b)) shows high concentration around the QD and rapid attenuation outwards, indicating preferential coupling to the nearest QD.

Using core/shell QDs as the basis for the waveguide, an earlier model was developed to calculate the pump power necessary to achieve net optical gain by solving the generation and recombination rate equations given the condition of charge neutrality. With a 5 nm × 5 nm × 5 nm CdSe/ZnS core/shell structure, it was found that net linear gain developed between 10 to 100 pW/QD pump power without considering the nonradiative Auger recombination effect. An ABCD matrix approach was previously employed to simulate the transmission efficiency for a 1D QD array, which confirms the tradeoff between interdot coupling efficiency and linear gain within each QD to achieve overall lossless waveguide transmission [29].

Due to the limitation of the current fabrication approach, in which the waveguide width is defined by electron-beam lithography (EBL), the actual waveguide will consist of a 2D array of QDs. In this section, a multipart theoretical model for estimating the throughput at variable gain levels in a 2D QD array is described. The nonradiative Auger recombination [34, 35] is incorporated into the previously developed gain model to determine the threshold pump intensity. Next, FDTD simulations provide a means of finding the interdot coupling efficiency relationship with respect to the position and distance of neighboring particles. Generalizing the two-quantum-dot problem to the large number of nanoparticles that form a device is accomplished through multiple-cycle Monte Carlo simulation using randomized QD placement. The results for 2D and 1D QD arrays reveal that the pump power/QD gain requirement for the 2D structure is in fact reduced. This prediction is confirmed by experimental measurements on waveguide loss coefficient, which will be shown in Section 11.3.

11.1.1
QD Gain vs. Pump Power

The optical gain response on pump power is found primarily through two equilibrium equations, one that balances absorption and emission processes and another that equates the number of holes in the valence band with electrons in the conduction band. The formulas are solved simultaneously for the two unknown electron and hole quasi-Fermi energies in the QD. Then, the gain, which is net emission subtracted by absorption as reflected by the difference between the quasi-Fermi levels, may be fully described given known material constants [29, 36]. The original model provides a foundation to estimate the threshold pump power for gain, optimal linear gain coefficient measured gain per pump power, and maximum gain achievable in the QD prior to saturation. However, Auger recombination, which is the dominant nonradiative competitive process in the QD, has been observed experimentally and must be included for a more comprehensive result.

To incorporate the Auger effect requires knowing the time constant associated with a particular material and size, which has been observed to follow the relation $\tau_A = \beta R^3$, where R is the particle radius and β is 5 ps/nm^3 for a QD composed of cadmium selenide. Moreover, the addition of a capping layer or a shell of zinc sulfide does not appear to affect the trend [34]. Using a sphere model to determine the radius necessary for a 655-nm emission QD leads to $R = 3.5$ nm, which results in $\tau_A = 214$ ps. The Auger recombination rate, r_A, may now be expressed as:

$$r_A = \frac{f_c(E_{1e})[1-f_v(E_{1h})]}{\tau_A} \int_{E_g}^{\infty} \frac{\hbar/\tau_{in}}{[E-(E_{1e}-E_{1h})]^2 + (\hbar/\tau_{in})^2} dE \quad (11.1)$$

where τ_{in} is the intraband relaxation time describing the broadening of the energy states due to electron-electron collisions, $f_c(E_{1e})$ and $f_v(E_{1h})$ are the Fermi distribution function in the conduction and valence bands at the first excited states, E_g is the bandgap energy and $\hbar$ is the reduced Planck's constant. Hence, (11.1) shows that the nonradiative recombination rate is weighted by the occupation of electrons and holes and has a finite energy distribution. The equilibrium rate equation can then be modified by adding the r_A term on the emission side:

$$r_{abs,02} = r_{st.ems,20} + r_{sp.ems,20} + r_{sp.ems,10} + r_A \quad (11.2)$$

The equation describes that in a three-level energy system (ground state, 1st excited state, and 2nd excited state), the absorption rate to the second excited state $r_{abs,0,2}$ must counterbalance the spontaneous relaxation process from the second $r_{sp.ems,20}$ and first states $r_{sp.ems,10}$ to ground as well as the stimulated emission rate between first and ground states $r_{st.ems,20}$ and the Auger recombination rate r_A.

With the revised (11.2) and the charge-neutrality equation, which remains the same, the quasi-Fermi energies and gain are recalculated. The outcome is an increase in the threshold pump power from 0.067 nW/QD to 7.8 nW/QD, although the maximum gain at saturation remains the same. The spectra, depicted in Figures 11.2(a) and (b), compare the original and revised modeling results to show that outside of

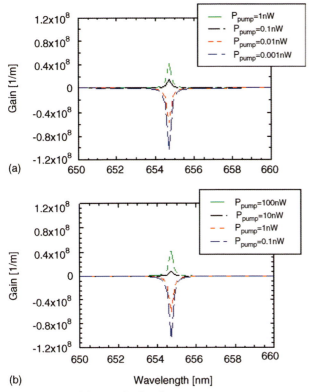

Figure 11.2 Modeling results for a linear gain spectrum as a function of pump power assuming (a) no Auger process and (b) with an Auger recombination lifetime of 214 ps.

the difference in required pump power to compensate for the nonradiative energy transfer, the trend in gain is identical. The result is expected since the Auger effect only impacts the efficiency of optical pumping, leaving the linewidth and the transition energy unchanged. The absorbed pump power necessary to achieve gain in each QD can be obtained by multiplying the absorption coefficient of the 405 nm pump wavelength at threshold, calculated to be 3.98×10^7 m^{-1}, with the particle diameter to find that $(1 - \exp(-3.98 \times 10^7$ m$^{-1} \times 7$ nm$)) = 0.244$ is the fractional absorption and 0.244×7.8 nW/QD gives a required pump power of 1.9 nW per QD.

From another perspective, the threshold optical intensity as influenced by Auger recombination may be defined as $I_{th} \approx \hbar\omega_p/(\sigma_{abs}\tau_A)$, where ω_p is the pump light frequency and σ_a is the absorption cross-section. The latter term, given by the formula $\sigma_{abs} = 2303\varepsilon_\lambda/N_A$, is dependent on the extinction coefficient, ε_λ, as measured in [(M cm)$^{-1}$] units and Avogadro's constant [37, 38]. From the manufacturer specification of the 655-nm emission CdSe/ZnS QD extinction ratio, we find $\sigma_{abs} = 2.14$ nm^2 at 405-nm excitation, which results in a threshold pump of 2.3 nW/QD and an intensity I_{th} of 1.07 mW/μm^2.

11.1.2
FDTD Modeling for Interdot Coupling

In addition to the optical gain in the QD, the near-field coupling efficiency between adjacent QDs is another critical factor for the overall waveguide transmission efficiency. Finite-difference time-domain (FDTD) simulation is used for determining the near-field intensity distribution of the stimulated emitted photons as a starting point and OptiFDTD v7 package from OptiWave [39] is employed as the simulation tool. The QD is modeled as an 8-nm diameter dielectric sphere at the center of a 20 nm × 20 nm × 20 nm simulation volume, embedded with a directional point source representing the stimulated photon emission. The latter is represented by a 655-nm wavelength line source placed at the origin of the QD and specified with a 0.4-nm full width at half-maximum (FWHM) profile directed to propagate forward along the waveguide direction (z-axis). While the 0.4-nm source leads to higher divergence with respect to distance from the QD, choosing a wider FWHM would result in reduced divergence and lead to a less conservative estimate for interdot coupling.

Through boundary-condition matching, Maxwell's equations for electric and magnetic fields are solved iteratively across the sample space and over time until a converged solution emerges. The mesh or border size in the simulation volume is a user-defined variable, which trades accuracy of the electromagnetic behavior for computational speed to arrive at a result. For current simulation conditions, 5000 time steps of 2.508×10^{-19} s and 3D solutions are chosen to find all the x, y, and z components of the electric and magnetic fields. Additionally, the boundary conditions are perfect match layers in x, y, z and the polarization of the source is linear in the y (height) direction. The coupling efficiency is estimated as the ratio of the intensity flow (represented by the Poynting vector) integrated over the adjacent QD cross-sectional area compared to the integration over all space:

$$\eta = \frac{P_{absorb}}{P_{total}} = \frac{\int \vec{S} \cdot \vec{da}}{\int \vec{S} \cdot \vec{dA}} \tag{11.3}$$

To optimize for the convergence time using mesh size, η was found for mesh sizes between 2 nm to 0.15 nm, such that the cell size varies from 2 nm × 2 nm × 2 nm to 0.15 nm × 0.15 nm × 0.15 nm, respectively. The simulation results showed a plateau in the coupling efficiency at the smaller mesh size, confirming the convergence. Figure 11.3(a) maps the resulting forward η with respect to interdot distance measured from edge to edge of the QD. The curve shows the trend of increasing coupling efficiency to 0.5 at a short distance outside of the QD due to the spatial distribution of the electric field (see Figure 11.1(b)) [33] and then diminishes at farther separations. On the other hand, Figure 11.3(b) demonstrates the cross-coupling in the $+\hat{x}$ and $-\hat{x}$ directions (along the transverse direction of the waveguide), which is indicative of a crosstalk component. The latter is a negative value as the energy is flowing down or out of the QD along the negative portion of the y-axis. The slight asymmetry is attributed to the offset in mesh divisions on the xyz-axes over the entire volume.

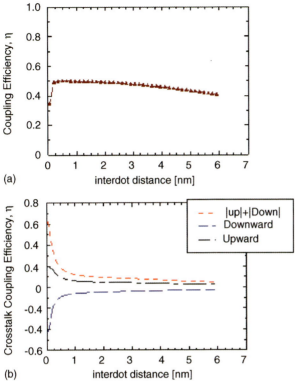

Figure 11.3 FDTD simulation of interdot coupling efficiency versus dot separation: (a) Coupling in propagating direction. (b) Cross-coupling in upward and downward (transverse) directions.

Although Förster [40] and exciton–polariton coupling [41, 42] provide different energy-transfer mechanisms in the near-field, neither is sensitive to the directional aspect of photons in stimulated emission. Instead, the first method is based solely on an overlap between spontaneous emission and absorption spectra, while the second arises from the effect of the optical field emitted by the QD into the environment to form exciton–polaritons and coupled to adjacent absorbers. Therefore, the FDTD-based method represents a more accurate approach to determining coupling efficiency in the QD waveguide case.

11.1.3
Monte Carlo Simulation for Transmission Efficiency

With the interdot coupling efficiency as a function of QD separation determined, the transmission characteristics for a system of QDs may now be calculated. Through Monte Carlo (MC) modeling, a population of QDs is randomly distributed in a two-dimensional array without overlapping. Then, throughput is found by successive calculation that considers the optical gain within and coupling between neighboring QDs as the signal travels along the propagation direction. By storing the relative output value from each QD waveguide formation over a large number of cycles, a statistical picture of the device behavior can be formed.

In order to implement the coupling efficiency in the MC simulation, curve fitting to the result shown in Figure 11.3(a) is first performed to give a R-squared value of 0.9993. Similarly, the same process for the transverse coupling efficiencies (Figure 11.3(b)) in terms of separation distance, d, result in exponential trends with:

$$\eta^{down} = 0.065 d^{-0.4564} \quad \text{and} \quad \eta^{up} = 0.0674 d^{-0.5537} \tag{11.4}$$

The R^2 values for (11.4) are 0.9668 for η^{up} and 0.9807 for η^{down}, confirming the close proximity of the function to the data. Apart from the fitting function, a waveguide length, width, QD diameter, D, and maximum interdot distance, d_{max}, are the remaining specifications for the MC simulation.

Defining the z- and x-axes to represent length and width directions, the y or height component is a fixed magnitude given by the QD diameter across the entire structure. Then, the signal output at the exiting face of each QD is approximated by the sum of three parts, which includes the forward coupling as well as the cross-coupling from QDs above and below the one of interest, multiplied by the optical gain effect through the QD:

$$T_{i,j} = e^{G\,D} \left(T_{i-1,j} \eta_{i,j} W_i + \sum_{k \leq i} T_{k,j-1} \eta^{up}_{j-1} L_k + \sum_{k \leq i} T_{k,j+1} \eta^{down}_{j+1} L_k \right) \tag{11.5}$$

The indices i and j denote the QD's z and x position within the randomized array. In (11.5), summations are used for the cross-coupled terms to account for all relevant neighbors, where L and W act as the overlap coefficients in the lateral and propagation directions that decrease linearly with the x and z position offset between the original and adjacent particles. Equation (11.5) represents a lower bound to the estimation as all the diagonal couplings along the propagation are represented by the cross-coupling terms ($k<i$, η^{up}, and η^{down}), which in general have a lower coupling efficiency than the forward coupling. Finally, the output at each point in the chain is weighted by the quantum-yield value, $Q_{YD} = 0.86$, as provided by the manufacturer to describe the emission to absorption efficiency.

Equation (11.5) needs to be applied recursively to simulate the overall propagation through the QD chain. For a 2D QD array of 2 µm length and 500 nm width, which corresponds to 155 by 40 particles with $d_{max} = 10$ nm, the median result for 500 simulations is shown in Figure 11.4(a), which illustrates the dependence on the optical-gain coefficient through the QD. Over the range of 1×10^7 to 5×10^7 m^{-1} linear gain coefficient, the relative transmission swings from 10^{-6} to 10^8. The lossless point settles at 3.1×10^7 m^{-1}, which is below the saturation point provided by the QD gain calculation result (Figure 11.2(b)) and represents an attainable amount.

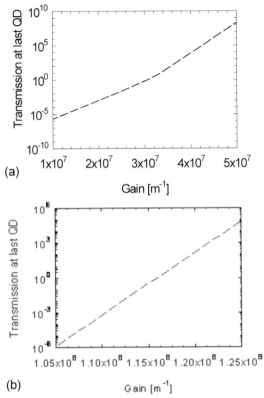

Figure 11.4 Monte Carlo simulation of gain-dependent QD waveguide transmission with randomized interdot separations and FDTD-determined coupling profile. The simulation is performed and averaged over 500 cycles. (a) 2D QD array result with 155 × 40 particles corresponding to a 500-nm wide, 2-μm long waveguide. (b) 1D QD array result with 155 particles corresponding to a 2-μm long waveguide.

To compare the 2D QD array case with the 1D case, we can reformulate the output at each QD simply as:

$$T_i = e^{GD} \cdot T_{i-1} \cdot \eta_i \tag{11.6}$$

where only gain, G, and coupling, η, from the previous particle is relevant. Figure 11.4(b) depicts the gain-dependent median throughput after adding the quantum-yield effect for 500 simulation cycles of a waveguide with 2 μm length (155 particles), $d_{max} = 10$ nm and $D = 8$ nm. Unity transmission occurs at ∼11.6 × 10^7 m^{-1} gain, which is almost four times higher than the 2D array. The difference demonstrates that a wider waveguide reduces the gain threshold for lossless throughput due to the cross-coupling, at the expense of increased waveguide width. Therefore, depending on the design criteria, array width and desired output amplitude may act as tradeoff variables.

11.2
From Concept to Realization – Fabrication of the Device

Two molecular self-assembly fabrication processes have been explored to fabricate the QD waveguides on a silicon substrate. They are mediated through either 3′mercaptopropyl–trimethoxylsilane (MPTMS) followed by DNA or 3′aminopropyltriethoxylsilane (APTES). The fundamental steps are similar in both processes as electron-beam lithography (EBL) or high-resolution optical lithography defines the overall waveguide pattern and the QDs are ultimately bound to the substrate via a series of chemical reactions. While the former offers programmability through DNA-specific binding, the latter provides an efficient way to fabricate the device with high QD coverage.

11.2.1
DNA-Directed Self-Assembly Fabrication

Desoxyribonucleic acid (DNA) has strong potential as a material to aid in the assembly of nanostructures. It has been demonstrated to assemble nanoparticles [43–46], or to direct attachment of nanoparticles onto micrometer-scale features [47, 48]. Anchoring single-stranded DNA (ss-DNA) to the surface of a substrate establishes highly discriminatory binding sites, as only reaction with the complementary chain (cDNA) will allow for further, sequential assembly steps. The presence of noncomplementary strands does not lead to hybridization. Consequently, by depositing ss-DNA at specific locations, we can program the eventual assembly of quantum dots. The benefit becomes apparent for making multiple-type QD waveguides.

11.2.1.1 Self-Assembly Process and Characterization
The process flow to assemble QDs through DNA mediation is illustrated in Figure 11.5 [29, 36]. An oxidized silicon wafer is chosen as the substrate. In principle, any substrate covered with oxidized material can be used as the starting substrate. Cleaving a small coupon from a wafer, the sample is first cleaned with xylene, acetone, isoproypl alcohol (IPA) and de-ionized (DI) water and blown dry with a nitrogen stream. A 3% dilution of PMMA is spin coated on the surface to form a layer ~90 nm thick. After a prebake at 180 °C for 90 s, the piece is transferred to the SEM vacuum chamber for EBL. Afterwards, the sample is developed in 1 : 3 methyl isobutyl ketone (MIBK):IPA for 70 s, rinsed with IPA and N_2 dried. At this point, diffraction from the waveguide-pattern trenches may be observed under the optical microscope as confirmation of the e-beam writing.

The sample is then treated with oxygen plasma to generate hydroxyl (–OH) groups, which covalently bind to the dangling ends of silicon oxide and act as a hook for the next chemical reaction. Next, the first self-assembled monolayer (SAM) is formed by gas-phase deposition of an organo-silane molecule, 3′mercaptotrimethoxysilane (MPTMS). The silane terminal anchors to the oxygen element of the hydroxyl groups, thus freeing the hydrogen as an energy-minimization step, while the mercapto, or thiol (SH) group becomes oriented to the outer surface. After a period of two hours in

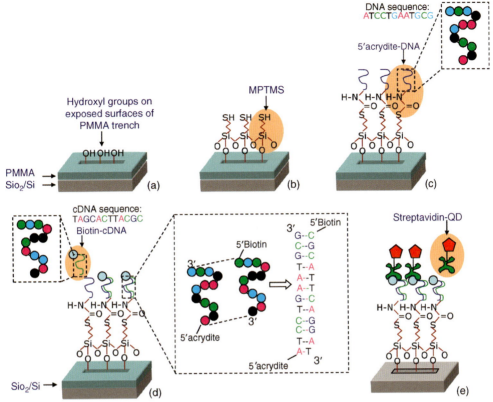

Figure 11.5 DNA-directed QD waveguide fabrication process: (a) pattern waveguides on PMMA-coated substrate by EBL and treat with oxygen plasma, (b) deposit MPTMS monolayer, (c) covalently bind with 5′acrydite-DNA, (d) hybridize with biotin-modified cDNA, (e) bind streptavidin-QDs to biotin-cDNA sites; remove PMMA [29].

which the sample and MPTMS co-mingle, vacuum is turned on, evacuating the chamber for one hour to remove excess MPTMS. Then, the sample is removed and rinsed with IPA, blown with N_2, and placed on the hot plate set to 80 °C for 10 min. The latter steps are implemented to help with the formation of a MPTMS mono- rather than multilayer [49, 50]. Verification of the MPTMS monolayer can be achieved using XPS, which measures the elemental structure within 1–10 nm of a surface. The samples with MPTMS monolayer yield average % compositions of 58.6% oxygen, 28.3% silicon, 12.4% carbon, and 0.7% sulfur. The untreated or control samples demonstrate no sulfur content with 63.0% oxygen, 29.6% silicon, and 7.4% carbon. The carbon presence is due to the MPTMS chemical structure as well as finite environmental contamination.

The second monolayer, composed of a 12-base pair (5′acrydite-ATCCTGAATGCG-3′) chain of DNA terminated by acrylamide, is deposited on the sample in solution form [51]. The base DNA is diluted to 10 µM concentration such that a 500-µL volume is composed of 440 µL DI, 50 µL 1× phosphate buffer solution (PBS), 5 µL 3 mM $MgCl_2$ and 5 µL of 1 mM of the acrydite-DNA. The sample is left to interact with base DNA overnight and is subsequently rinsed with 1× PBS and blown with N_2. Immersion in buffered acrylic acid followed by PBS rinse and N_2 dry leads to passivation of inactivated MPTMS molecules and readies the sample for addition of the complementary DNA (cDNA).

Biotin-linked cDNA is used to establish the third SAM with the same deposition procedure. After this step, the sample is ready for QD deposition. Streptavidin-bound QDs are dispensed in 1× PBS to about 0.1 µM concentration and dropped onto the surface. The quantum-dot assembly step utilizes the well-known biological recognition avidin–biotin binding pair where streptavidin, which is a protein, has four sites that preferentially attract and bind to biotin molecules. In the fabrication process, the sample is treated with QD solution for 30 min and rinsed and immersed for 5 min for three times with 2× SSPE (3.0 M sodium chloride, 0.2 M sodium hydrogen phosphate, 0.02 M EDTA, pH 7.4). The last step is to remove the PMMA layer using a 3-min toluene submersion, rinse with IPA and dry the sample with N_2 gas.

Figure 11.6 shows the fluorescence image of QDs patterned with the described DNA-directed self-assembly process. The peak wavelength of the QD emission spectrum is 655 nm. QD waveguides with different widths and lengths are fabricated following the same procedure. The SEM image in Figure 11.7(a) shows a marked boundary inside and outside the e-beam patterned area with strict QD adherence within the waveguide region. The fluorescence imaging for some waveguides between 0.1 to 5 µm widths, in Figure 11.7(b), shows some PMMA residue, which may be reduced by changing the solvent. A likely candidate is dichloromethane, which is used in the succeeding APTES-mediated self-assembly method. Moreover, QD coverage may be improved with increasing SAM solution concentrations and sample agitation

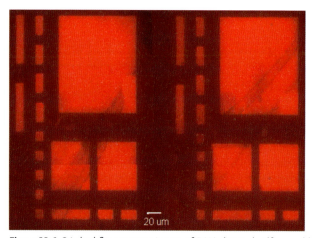

Figure 11.6 Stitched fluorescence image of DNA-directed self-assembled QD pattern.

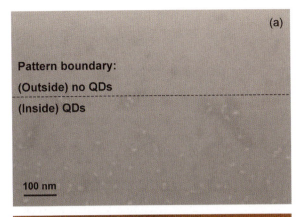
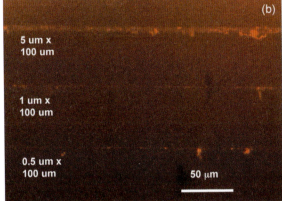

Figure 11.7 (a) SEM image demonstrating QD deposition confined within the waveguide region. (b) Fluorescence image of 5 µm, 1 µm and 0.5 µm width QD waveguides [52].

during the deposition steps. From a different perspective, the PMMA layer may be thinned through additional dilution or higher spin speed to create shallower waveguide pattern trenches. This modification would allow molecules to better enter into the patterned area and assemble. On the whole, the results indicate the feasibility of DNA-directed fabrication of QD waveguides, encompassing six successive chemical interactions. The essential benefit of the method is that of DNA sequence-based programmable assembly, which may be exploited for multiple QD-type structures.

11.2.1.2 Programmable DNA-Directed Self-Assembly

While single QD-type assembly is the basis for device fabrication, in order to augment component diversity and complexity, we need a way to mass produce multiple patterns using different kinds of nanoparticles. As streptavidin conjugated structures will preferentially bind to biotin, the critical step is to prepare the QDs with the corresponding biotin-complementary DNA molecules before reacting them en masse to the base DNA on the substrate. The details of the procedure are much like before, but a few additional steps are required.

A sequence of EBL is needed to define each of the regions upon which different QDs will be attached. After each EBL step, the substrate is treated with MPTMS followed by the first acrydite-terminated DNA. With adequate concentration and exposure time, the packing of the base layer with quench molecules will be dense and prevent assembly of subsequent DNA strands. Since the process of stripping and spin coating a fresh layer of PMMA may harm the biological components on the surface, the original PMMA layer is reused to ensure that the first DNA SAM does not further react. The next EBL step creates new patterns and the second round of MPTMS and different acrydite-DNA sequence SAMs are added. The process is repeated for as many QD types, and hence DNA–cDNA pairings, that are desired.

After the final set of base DNA is formed on the substrate and prior to assembly with the complementary strands, different biotin–cDNA chains are mixed with the matching QD solutions specific to the device design. Subsequently, the QD–cDNA solutions are deposited on the sample as a mass self-assembly step. Figure 11.8 illustrates the process that takes advantage of DNA hybridization selectivity to simultaneously form a diverse range of QD structures. As is customary, there is a 30-min period for the base pairs to bind. Gentle agitation may improve the reaction rate and homogenize the coverage. Finally, the sample is rinsed and immersed for 10 min in $2\times$ SSPE, rinsed in 0.3 M ammonium acetate and N_2 dried. The PMMA is stripped with toluene or dichloromethane, washed with DI water and again N_2 dried. The benefit of using DNA is fully exploited by the multiple QD-type fabrication. Such capability gives rise to diversity in device manufacture. Patterning of wavelength-selective waveguides, photodetectors, filters are within reach and developing the process is important to making QD-based nanophotonic circuits.

11.2.2
Self-Assembly Through APTES

If programmability is not required, fabrication of the QD waveguide can be achieved using a simple two-layer molecular self-assembly process mediated through 3′aminopropyltri ethoxysilane (APTES).

11.2.2.1 Self-Assembly Process and Characterization
Similar to the DNA-assisted technique, the APTES method (Figure 11.9) starts with oxygen plasma treatment at 20 W for 60 s following EBL of the waveguide pattern. Then, the process diverges with sample immersion into 0.1–0.2% v/v APTES thoroughly mixed in a 95% IPA and 5% DI H_2O solvent; for instance, 4 μL APTES would be mixed in 1900 μL IPA and 100 μL DI to make 0.2% solution. After 1 min reaction time, the sample is removed and rinsed in IPA. In contrast to MPTMS, the chemisorption of APTES leads to an amine ($-NH_3$) rather than a mercapto ($-SH$) group presented at the terminal. Afterwards, the sample is placed on a hot plate for 7.5 min at 110 °C to cure the monolayer [53].

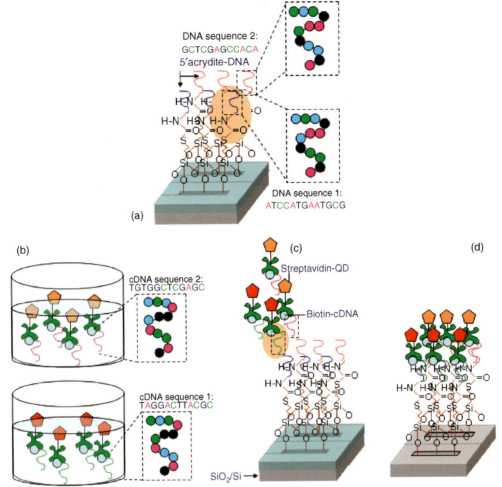

Figure 11.8 Fabrication of multiple QD-type waveguides, post EBL of the second set of waveguide patterns and MPTMS SAM deposition: (a) covalently bind with 5'acrydite-DNA #2, (b) bind different streptavidin-QDs with corresponding biotin-modified cDNA, (c) drop mixed solutions and allow hybridization of cDNA #1 with DNA #1 and cDNA #2 with DNA #2, (d) remove PMMA [52].

Similar to MPTMS, deposition of the APTES SAM can be confirmed with XPS, which finds existing nitrogen content compared to none for the controls. The average % compositions are 1.2% nitrogen, 29.8% silicon, 54.5% oxygen, and 14.4% carbon while the control gave an average of 60.6% oxygen, 32.3% silicon, and 7.1% carbon. Additional confirmation is made with AFM, in which a step height of 0.8 nm, equivalent to one monolayer, occurs in the waveguide pattern given that the PMMA layer is removed at this stage. Continuing on, a droplet of 0.125 µM carboxylated

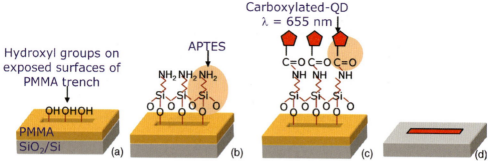

Figure 11.9 QD waveguide self-assembly fabrication through APTES: (a) EBL pattern a SiO$_2$/Si sample coated with PMMA; develop PMMA; treat with oxygen plasma to create hydroxyl groups on surface; (b) Solution phase deposition of APTES; (c) Carboxylated QDs covalently bind to the amine groups of APTES; (d) Strip PMMA to leave waveguide [30].

QDs mixed in DI H$_2$O with 1 mM 1-ethyl-3-(3′dimethyl-aminopropyl)-carbodiimide (EDC), a coupling reagent that aids in binding amine and carboxyl groups, is deposited directly on the patterned region. After one hour wait time, the QD solution is rinsed off the substrate with 1× phosphate buffer solution followed by 0.3 M ammonium acetate to remove the salts from the buffer. The sample is then blown dry with N$_2$. For the final steps, the PMMA is dissolved with dichloromethane, CH$_2$Cl$_2$, via a three-minute immersion process [54], and the sample is rinsed with DI H$_2$O and again dried under a nitrogen stream.

QD assembly as waveguides is determined by fluorescence and AFM, where distinct emission and topological changes are observed. Figures 11.10 and 11.11 depict 500-nm and 100-nm wide waveguides defined by QDs emitting at 655 nm with consistent coverage. The height profile, which reflects the combined APTES and QD layers, is about 10 nm.

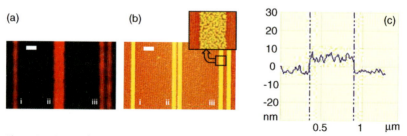

Figure 11.10 (a) Fluorescence and (b) AFM images of 500-nm wide waveguide: (i) single; two adjacent waveguides spaced (ii) 200 nm apart and (iii) 500 nm apart [30], scale bar is 1 μm; (c) section height of the 500-nm wide waveguide determined by AFM.

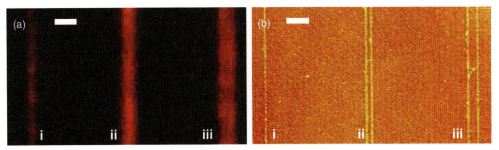

Figure 11.11 (a) Fluorescence and (b) AFM images of 100-nm wide waveguide, resolution is diffraction limited: i) single; two adjacent waveguides spaced ii) 200 nm apart and iii) 500 nm apart [30], scale bar is 1 µm.

11.2.2.2 Multiple-QD-Type Waveguide Fabrication

By repeating the fabrication process of the APTES-guided self-assembly, it is possible to create waveguides using QDs of different emission wavelengths to operate at targeted transmission wavelengths. The fabrication process is depicted in Figure 11.12. After self-assembly of the first QD waveguides, the PMMA layer is

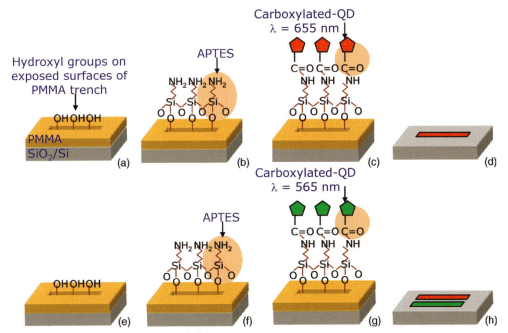

Figure 11.12 Multiple QD-type waveguide fabrication process: (a)–(d) directly follow single QD type process, then a fresh layer of PMMA is spin coated onto the sample, and the fabrication cycle is repeated in (e)–(h). Use of alignment markers in EBL enables accurate spacing between multiple waveguide patterns [52].

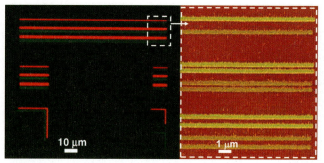

Figure 11.13 Fluorescence micrograph (left) and AFM of zoomed-in region (right) confirming multiple QD-type waveguide deposition. The emission wavelengths are 655 nm and 565 nm, corresponding to larger and smaller QDs. The different colors in the AFM image demonstrate the height difference between the two types of QDs [52].

removed and a fresh PMMA layer is deposited for the fabrication cycle of the second QD waveguides. There may be some concern about whether spin coating of a new PMMA resist would dislocate the QDs and adversely affect the existing devices via the 4500 rpm spinning process. Likewise, the small, but non-negligible profile of the assembled waveguides may create some features in the new PMMA layer to affect the e-beam distribution. However, fabrication results proved the two issues as insignificant.

Figure 11.13 shows the fluorescence and topological profile for 655-nm and 565-nm emission QD waveguides, which were formed directly following the fabrication process described in Figure 11.12. There is even deposition for both types of QDs and single monolayers are formed. The height is clearly greater for the red QDs, corresponding to larger nanoparticles, as demonstrated by the brighter coloring consistent within the confines of the line. Through the aid of alignment markers for EBL, accurate positioning of the waveguide patterns and therefore control of spacing between QD waveguides can be achieved.

11.2.3
Discussion on Fabrication Methods

As detailed in this section, QD waveguide fabrication has been realized by two major methods, namely DNA-directed and APTES-guided molecular self-assembly. The former technique provides a way for programmable deposition; however, more work would be required to improve the density of QD coverage. Optimization of each layer, namely MPTMS, acrydite-DNA, biotinylated cDNA, streptavidin-conjugated QDs, and the chemical interaction between the layers would help build out the process step by step. On the other hand, the two-layer molecular self-assembly approach using APTES coupled to carboxylated QDs allows for quick fabrication turn-around of several hours with good nanoparticle homogeneity and

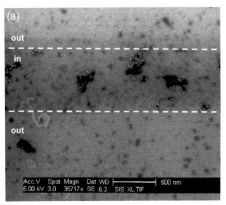 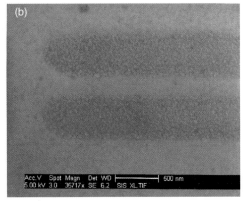

Figure 11.14 SEM images of 500-nm wide waveguides formed by (a) DNA-directed and (b) APTES-guided self-assembly methods [52].

demonstration of multiple QD type structures. Figure 11.14 compares the high-resolution SEM images of two 500 nm-wide waveguides formed by both procedures with clear display of the high coverage using APTES-mediated self-assembly. On the whole, the key points to consider are the tradeoffs in transitioning from the DNA-directed to APTES-guided technique. While faster progress and time to test are achieved with the latter, the programmable element is missing. However, the multi-QD-type proof-of-principle structures offer an alternative path to fabricating various wavelength-specific devices.

11.3
How Well the Devices Work – A First Probe

To test the QD waveguide, input and output to the device are coupled through tapered optical fibers. The 639-nm input laser signal is modulated by an optical chopper, and the output signal measured by a femto-watt photodetector is sent to a lock-in amplifier. The nonmodulated 405-nm pump source and spontaneous emission from the QDs can therefore be filtered out using this setup, and only the stimulated emission initiated by the input signal is detected. Both the pump light and the signal light power levels are monitored by optical-fiber couplers to calibrate the measurement results. A customized computer data-acquisition and control program allows for the specification of signal and pump settings as well as the duration of each test and time between measurements. Each data series is a compilation over each laser setting of the output lock-in voltage or optical power with the tapped signal and pump values as available. With the files, MATLAB is used to perform analysis to extract the device performance. In particular, the measurements are averaged over the same laser source parameters. Generally, the standard deviation is two orders of magnitude smaller than the mean values.

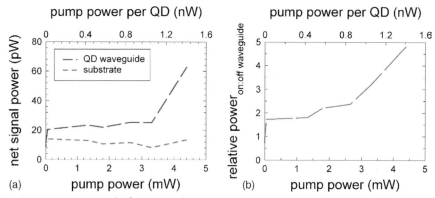

Figure 11.15 Test results for a 10-μm long, 500-nm wide QD waveguide. (a) Net transmitted signal power, and (b) relative power ratio with and without the waveguide.

11.3.1
Waveguiding with Flexibility

Verification of the QD array as a waveguiding device can be carried out by comparing the transmitted signal with and without the waveguide (direct fiber-to-fiber coupling on the surface of the substrate at the same distance). The experimental results in Figure 11.15 show the net signal data and relative transmission measured from a 500-nm wide, 10-μm long straight waveguide [30]. The curves following in Figure 11.16 are the results for a 500-nm wide, 20-μm long waveguide with 90° bend in the middle. The APTES-mediated two-layer molecular self-assembly

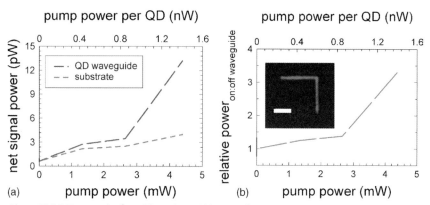

Figure 11.16 Test results for a 20-μm long, 500-nm wide QD waveguide with 90° bend in the middle. (a) Net transmitted signal power, and (b) relative power ratio with and without the waveguide; inset: fluorescence micrograph of the device, the scale bar is 5 μm.

fabrication process is used to create these test devices as QD deposition density is much higher than the DNA-directed method.

Rather than dielectric waveguiding where the optical mode largely travels within a region of high refractive index embedded in a lower-index material, the QD waveguide operates through a series of photons cascading downstream through stimulated emission processes. The experiment shows that the net transmitted signal power with the waveguide is always higher and the ratio increases with the pump power, confirming the waveguiding mechanism of the device. A point to note is that the 90°-bend waveguide exhibits similar waveguiding behavior with slightly lower ratio. Given that the cross-coupling to the lateral direction is significant at small QD separation, observed from the near-field intensity distribution in Figure 11.1(b) and the FDTD simulation results in Figure 11.3, such results are reasonable. The reduced power ratio for the 90°-bend waveguide may stem from the fact that it is twice as long as the straight waveguide. The capability for guiding through sharp bends, which cannot be easily attained by conventional dielectric waveguides, is a desirable feature for high-density photonic circuitry with flexible routing.

11.3.2
Loss Characterization

Extending the waveguide-characterization procedure across multiple lengths allows us to determine the loss coefficient. In loss quantification, QD waveguides from 4 μm to 10 μm length at 500 nm (Figure 11.17(a)) and 100 nm (Figure 11.17(b)) widths were examined. Varying the pump power per QD from 1.18 to 2.08 nW produced the same upward trend in transmission at higher optical intensities. Additionally, the narrow waveguides revealed steeper slopes. An exponential fit to the data is used to determine the loss coefficient since the output power follows the relation, $P(z) = P_0 \exp(-\alpha z)$, with respect to the input power P_0 and the loss coefficient, α. As a result, an average loss value of 3 dB per 2.26 μm and 4.06 μm for the 100 nm and 500 nm width waveguides across the three pump settings are determined. The lower loss for

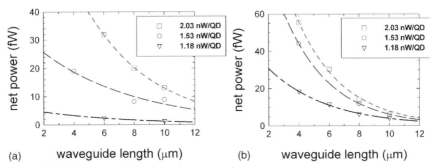

Figure 11.17 QD waveguide loss behavior measured over multiple lengths and pump powers. (a) 500-nm wide waveguides. (b) 100-nm wide waveguides.

wider QD waveguides confirms the MC simulation results shown in Section 11.1.3. Table 11.1 summarizes the reported theoretical and experimental findings for all proposed subdiffraction waveguiding methods.

11.4
To Probe Further – Summary and Outlook

With the versatility of self-assembly chemistry and near-field energy-transfer properties of semiconductor nanoparticles, the QD waveguide presents a possible route for subdiffraction-limit optical-energy transmission. In the chapter, we first described the operating principle and presented simulation of the device. The theoretical analysis consists of three parts: (1) Optical gain in the QD using equilibrium rate and carrier concentration equations. (2) Near-field energy-transfer efficiency utilizing FDTD simulations. (3) Waveguide transmission efficiency through a Monte Carlo approach, simulating an array of randomly distributed QDs.

Once the principle of operation was established, two molecular self-assembly fabrication methods were explored. The first one is a DNA-directed process that offers programmability, the second one is an APTES-mediated two-layer process that provides efficient fabrication with good QD coverage in the waveguide. A device fabricated using the latter method was tested. The measurement results demonstrated comparable waveguiding behavior through a straight waveguide and a waveguide with 90° bend. The loss coefficient was also characterized by measuring the output signal power of QD waveguides with various lengths. The extracted values of 3 dB/4.06 μm for 500-nm wide and 3 dB/2.26 μm for 100-nm wide waveguides show an improvement of throughput when compared to prior art subdiffraction-limit waveguiding structures using negative-dielectric materials, and confirm the trend shown by the Monte Carlo simulation.

While the basic principle and operation of the QD waveguide has been demonstrated, there is certainly room for improvement. Creating a lossless device for subdiffraction-limit optical-energy transmission would be ideal for ultracompact photonic integrated circuits. As the modeling results indicate, increasing gain may compensate for interdot coupling inefficiency. Moreover, the required gain levels are attainable prior to saturation from the finite energy states in the QD. What remains is that the pump requirement for the gain must be feasible for the system. With the current colloidal QDs, high pump power is necessary to form biexcitons for stimulated emission given the significant effect of Auger recombination. However, recent work on colloidal quantum wells [58, 59] and type-II QDs, where the core/shell structure is inverted to tailor the bandgaps for separation of the electron and hole wavefunctions [60], makes possible single-exciton gain at lower threshold energy to circumvent the Auger process and enable sub-nW/QD threshold pump power. Adapting the process to the modified QDs would advance the waveguide performance towards an amplified response and create more opportunities to form gain-enhanced devices at the nanoscale. Regardless, the promise of fabricating the colloidal QDs on a versatility of substrates through self-assembly and the favorable loss coefficient results demonstrate the QD array as an effective component for the nanophotonics toolbox.

Table 11.1 Summary of theoretical and experimental loss for subdiffraction waveguiding methods.

Method	Device dimensions	Wavelength	Theoretical loss	Experimental loss
Ag pin 1D fiber [55]	20-nm diameter core	633 nm	7.31 dB/μm =3 dB/410 nm	N/A
Au nanowire [16]	200 nm width 50 nm thickness	800 nm	N/A	1.7 dB/μm =3 dB/1.76 μm
Ag wedge [18]	300 nm base 40° angle	632 nm	1.9 dB/μm =3 dB/1.58 μm	2.9 dB/μm =3 dB/1.03 μm
Ag nanoparticle array [26]	50-nm diameter 25 nm interdot separation	488 nm	4.8 dB/μm =3 dB/614 nm	N/A
Ag nanoparticle array [13]	50 nm diameter 50 nm interdot separation	570 nm	30 dB/μm =3 dB/100 nm	31 dB/μm =3 dB/97 nm
Au IMI [56]	2D coverage 45 nm thickness	1.55 μm	0.00076 dB/μm =3 dB/3.9 mm	N/A
Au IMI in Si [20]	150 nm width 250 nm thickness	1.55 μm	0.55 dB/μm =3 dB/5.45 μm	0.8 dB/μm =3 dB/3.75 μm
Au clad MIM [57]	150 nm width 100 nm thickness	633 nm	12.2 dB/μm =3 dB/246 nm	N/A
Au-clad index-guided MIM [57]	150 nm width 100 nm thickness	633 nm	3.1 dB/μm =3 dB/968 nm	N/A
QD waveguide	500 nm width 100 nm width	639 nm	N/A	3 dB/4.06 μm 3 dB/2.26 μm

References

1 Johnson, S.G., Villeneuve, P.R., Fan, S. and Joannopoulos, J.D. (2000) Linear waveguides in photonic-crystal slabs. *Physical Review B-Condensed Matter*, **62**, 8212–8222.
2 Yanik, M.F., Fan, S., Soljacic, M. and Joannopoulos, J.D. (2003) All-optical transistor action with bistable switching in a photonic crystal cross-waveguide geometry. *Optics Letters*, **28**, 2506–2508.
3 Yablonovitch, E. (2003) Photonic bandgap based designs for nano-photonic integrated circuits. *International Semiconductor Device Research Symposium*, p. 300.
4 Bogaerts, W., Baets, R., Dumon, P., Wiaux, V., Beckx, S., Taillaert, D., Luyssaert, B., Van Campenhout, J., Bienstman, P. and Van Thourhout, D. (2005) Nanophotonic waveguides in silicon-on-insulator fabricated with CMOS technology. *Journal of Lightwave Technology*, **23**, 401–412.
5 Lipson, M. (2004) Overcoming the limitations of microelectronics using Si nanophotonics: Solving the coupling, modulation and switching challenges. *Nanotechnology*, **15**, S622–S627.
6 Xu, Q., Almeida, V.R., Panepucci, R. and Lipson, M. (2004) Experimental demonstration of guiding and confining light in nanometer-size low-refractive index material. *Optics Letters*, **29**, 1626–1628.
7 Barrelet, C.J., Greytak, A.B. and Lieber, C.M. (2004) Nanowire photonic circuit elements. *Nano Letters*, **4**, 1981–1985.
8 Law, M., Goldberger, J. and Yang, P. (2004) Semiconductor nanowires and nanotubes. *Annual Review of Materials Research*, **34**, 83–122.
9 Law, M., Sirbuly, D.J., Johnson, J.C., Goldberger, J., Saykally, R.J. and Yang, P. (2004) Nanoribbon waveguides for subwavelength photonics integration. *Science*, **305**, 1269–1273.
10 Dionne, J.A., Sweatlock, L.A. and Atwater, H.A. (2006) Highly confined photon transport in subwavelength metallic slot waveguides. *Nano Letters*, **6**, 1928–1932.
11 Zia, R., Schuller, J.A. and Brongersma, M.L. (2006) Near-field characterization of guided polariton propagation and cutoff in surface plasmon waveguides. *Physical Review B-Condensed Matter*, **74**, 165415.
12 Bozhevolnyi, S.I., Volkov, V.S., Devaux, E., Laluet, J.-Y. and Ebbesen, T.W. (2006) Channel plasmon subwavelength waveguide components including interferometers and ring resonators. *Nature*, **440**, 508–511.
13 Maier, S. (2006) Plasmonics: Metal nanostructures for subwavelength photonic devices. *IEEE Journal of Selected Topics in Quantum Electronics*, **12**, 1214–1220.
14 Takahara, J. and Kusunoki, F. (2007) Guiding and nanofocusing of two-dimensional optical beam for nanooptical integrated circuits *IEICE Transactions on Electronics*, **E90-C**, 87–94.
15 Dickson, R.M. and Lyon, L.A. (2000) Unidirectional plasmon propagation in metallic nanowires *The Journal of Physical Chemistry. B*, **104**, 6095–6098.
16 Krenn, J.R., Lamprecht, B., Ditlbacher, H., Schider, G., Salerno, M., Leitner, A. and Aussenegg, F.R. (2002) Non-diffraction-limited light transport by gold nanowires. *Europhysics Letters*, **60**, 663–669.
17 Yin, L., Vlasko-Vlasov, V.K., Pearson, J., Hiller, J.M., Hua, J., Welp, U., Brown, D.E. and Kimball, C.W. (2005) Subwavelength focusing and guiding of surface plasmons. *Nano Letters*, **5**, 1399–1402.
18 Pile, D.F.P., Ogawa, T., Gramotnev, D.K., Okamoto, T., Haraguchi, M., Fukui, M. and Masuo, S. (2005) Theoretical and experimental investigation of strongly localized plasmons on triangular metal wedges for subwavelength waveguiding. *Applied Physics Letters*, **87**, 061106:1–3.
19 Dionne, J.A., Sweatlock, L.A., Atwater, H.A. and Polman, A. (2006) Plasmon slot waveguides: Towards chip-scale propagation with subwavelength-scale

localization. *Physical Review B-Condensed Matter*, **73**, 035407.

20 Chen, L., Shakya, J. and Lipson, M. (2006) Subwavelength confinement in an integrated metal slot waveguide on silicon. *Optics Letters*, **31**, 2133–2135.

21 Brongersma, M.L., Hartman, J.W. and Atwater, H.A. (2000) Electromagnetic energy transfer and switching in nanoparticle chain arrays below the diffraction limit. *Physical Review B-Condensed Matter*, **62**, R16356–R16359.

22 Maier, S.A., Kik, P.G. and Atwater, H.A. (2003) Optical pulse propagation in metal nanoparticle chain waveguides. *Physical Review B-Condensed Matter*, **67**, 205402–205406.

23 Maier, S.A., Kik, P.G., Atwater, H.A., Meltzer, S., Harel, E., Koel, B.E. and Requicha, A.A.G. (2003) Local detection of electromagnetic energy transport below the diffraction limit in metal nanoparticle plasmon waveguides. *Nature Mater*, **2**, 229–232.

24 Sweatlock, L.A., Maier, S.A., Atwater, H.A., Penninkhof, J.J. and Polman, A. (2005) Highly confined electromagnetic fields in arrays of strongly coupled Ag nanoparticles. *Physical Review B-Condensed Matter*, **71**, 235408:1–7.

25 Maier, S.A., Barclay, P.E., Johnson, T.J., Friedman, M.D. and Painter, O. (2004) Low-loss fiber accessible plasmon waveguide for planar energy guiding and sensing. *Applied Physics Letters*, **84**, 3990–3992.

26 Quinten, M., Leitner, A., Krenn, J.R. and Aussenegg, F.R. (1998) Electromagnetic energy transport via linear chains of silver nanoparticles. *Optics Letters*, **23**, 1331–1333.

27 Klimov, V.I., Mikhailovsky, A.A., Xu, S., Malko, A., Hollingsworth, J.A., Leatherdale, C.A., Eisler, H.-J. and Bawendi, M.G. (2000) Optical gain and stimulated emission in nanocrystal quantum dots. *Science*, **290**, 314–317.

28 Chan, W.C.W. and Nie, S. (1998) Quantum dot bioconjugates for ultrasensitive nonisotopic detection. *Science*, **281**, 2016–2018.

29 Wang, C.-J., Lin, L.Y. and Parviz, B.A. (2005) Modeling and simulation for a nano-photonic quantum dot waveguide fabricated by DNA-directed self-assembly. *IEEE Journal of Quantum Electronics*, **11**, 500–509.

30 Wang, C.-J., Huang, L., Parviz, B.A. and Lin, L.Y. (2006) Sub-diffraction photon guidance by quantum dot cascades. *Nano Letters*, **6**, 2549–2553.

31 Wang, C.-J. and Lin, L.Y. (2007) Nanoscale waveguiding methods. *Nanoscale Research Letters*, **2**, 219–229.

32 Wang, C.-J., Parviz, B.A. and Lin, L.Y. (2008) Two dimensional array self-assembled quantum dot sub-diffraction waveguides with low loss and low crosstalk. *Nanotechnology*, **19** (29), 295201.

33 Huang, L., Wang, C.-J. and Lin, L.Y. (2007) A comparison of crosstalk effects between colloidal quantum dot waveguides and conventional waveguides. *Optics Letters*, **32**, 235–237.

34 Klimov, V.I., Mikhailovsky, A.A., McBranch, D.W., Leatherdale, C.A. and Bawendi, M.G. (2000) Quantization of multiparticle Auger rates in semiconductor quantum dots. *Science*, **287**, 1011–1013.

35 Wang, L.-W., Califano, M., Zunger, A. and Franceschetti, A. (2003) Pseudopotential theory of Auger processes in CdSe quantum dots. *Physical Review Letters*, **91**, 056404.

36 Wang, C.-J. (2004) Nanophotonic Waveguides by Self-assembled Quantum Dots *Master's Thesis, Electrical Engineering Department, University of Washington*.

37 Leatherdale, C.A., Woo, W.-K., Mikulec, F.V. and Bawendi, M.G. (2002) On the absorption cross-section of CdSe nanocrystal quantum dots. *The Journal of Physical Chemistry. B*, **106**, 7619–7622.

38 Klimov, V.I. (2000) Optical nonlinearities and ultrafast carrier dynamics in semiconductor nanocrystals. *The Journal of Physical Chemistry. B*, **104**, 6112–6123.

39 OptiWave Systems Inc. http://www.optiwave.com.

40 Förster, T. (1959) Excitation transfer. *Discussions of the Faraday Society*, **27**, 300–320.

41 Nomura, W., Yatsui, T., Kawazoe, T. and Ohtsu, M. (2007) The observation of dissipated optical energy transfer between CdSe quantum dots. *Journal of Nanophotonics*, **1**, 011591.

42 Sangu, S., Kobayashi, K. and Ohtsu, M. (2001) Optical near-fields as photon-matter interacting systems. *Journal of Microscopy*, **202**, 279–285.

43 Braun, E., Eichen, Y., Sivan, U. and Ben-Yoseph, G. (1998) DNA-templated assembly and electrode attachment of a conducting silver wire. *Nature*, **391**, 775–778.

44 Nykypanchuk, D., Maye, M.M., van der Lelie, D. and Gang, O. (2008) DNA-guided crystallization of colloidal nanoparticles. *Nature*, **451**, 549–552.

45 Mertig, M., Ciacchi, L.C., Seidel, R. and Pompe, W. (2002) DNA as a selective metallization template. *Nano Letters*, **2**, 841–844.

46 Park, S.Y., Lytton-Jean, A.K.R., Lee, B., Weigand, S., Schatz, G.C. and Mirkin, C.A. (2008) DNA-programmable nanoparticle crystallization. *Nature*, **451**, 553–556.

47 Mbindyo, J.K.N., Reiss, B.D., Martin, B.R., Keating, C.D., Natan, M.J. and Mallouk, T.E. (2001) DNA-directed assembly of gold nanowires on complementary surfaces. *Advanced Materials*, **13**, 249–254.

48 Sauthier, M.L., Carrol, R.L., Gorman, C.B. and Franzen, S. (2002) Nanoparticle layers assembled through DNA hybridization: characterization and optimization. *Langmuir*, **18**, 1825–1830.

49 Kurth, D.G. and Bein, T. (1993) Surface-reactions on thin-layers of silane coupling agents. *Langmuir*, **9**, 2965–2973.

50 Ramanath, G., Cui, G., Ganesan, P.G., Guo, X., Ellis, A.V., Stukowski, M., Vijavamohanan, K., Doppelt, P. and Lane, M. (2003) Self-assembled subnanolayers as interfacial adhesion enhancers and diffusion barriers for integrated circuits. *Applied Physics Letters*, **83**, 383–385.

51 Demers, L.M., Singer, D.S., Park, S.-J., Li, Z., Chung, S.-W. and Mirkin, C.A. (2002) Direct patterning of modified oligonucleotides on metals and insulators by dip-pen nanolithography. *Science*, **296**, 1836–1838.

52 Wang, C.-J. (2007) Sub-diffraction quantum dot nanophotonic waveguides PhD Dissertation, Electrical Engineering Department, University of Washington.

53 United Chemical Technologies, 'Using Silanes as Adhesion Promoters.'

54 Hu, W., Sarveswaran, K., Lieberman, M. and Bernstein, G.H. (2005) High-resolution electron beam lithography and DNA nano-patterning for molecular QCA. *IEEE Transactions on Nanotechnology*, **4**, 312–316.

55 Takahara, J., Yamagishi, S., Taki, H., Morimoto, A. and Kobayashi, T. (1997) Guiding of a one-dimensional optical beam with nanometer diameter. *Optics Letters*, **22**, 475–477.

56 Zia, R., Selker, M.D., Catrysse, P.B. and Brongersma, M.L. (2004) Geometries and materials for subwavelength surface plasmon modes. *Journal of the Optical Society of America A-Optics Image Science and Vision*, **21**, 2442–2446.

57 Kusunoki, F., Yotsuya, T., Takahara, J. and Kobayashi, T. (2005) Propagation properties of guided waves in index-guided two-dimensional optical waveguides. *Applied Physics Letters*, **86**, 21110.

58 Xu, J., Xiao, M., Battaglia, D. and Peng, X. (2005) Exciton radiative recombination in spherical CdS/CdSe/CdS quantum-well nanostructures. *Applied Physics Letters*, **87**, 043107.

59 Xu, J. and Xiao, M. (2005) Lasing action in colloidal CdS/CdSe/CdS quantum wells. *Applied Physics Letters*, **87**, 173117.

60 Klimov, V.I., Ivanov, S.A., Nanda, J., Achermann, M., Bezel, I., McGuire, J.A. and Piryatinski, A. (2007) Single-exciton optical gain in semiconductor nanocrystals. *Nature*, **447**, 441–446.

12
Hierarchy in Optical Near-fields and its Application to Nanofabrication

Makoto Naruse, Takashi Yatsui, Hirokazu Hori, Kokoro Kitamura, and Motoichi Ohtsu

12.1
Introduction

Nanophotonics has been showing great progress, exploiting the unique physical features appearing in light–matter interactions on the nanometer scale [1]. These interactions allow energy transfer and are free from the diffraction of light. High-density optical memories [2], sensors [3], and logic devices [4] have been demonstrated, as well as novel system functions [5] and nanoscale optical characterizations [6].

Fabrication techniques such as those based on lithographic and self-organization approaches are important assets in pursuing such nanophotonic devices and systems. Optical near-field interactions themselves have been successfully applied to fabrication techniques in the literature [7–10]; since an optical near-field appears locally in the vicinity of a material, it can help in fabricating nanostructures, such as in chemical vapor deposition (CVD) processes where materials are selectively deposited only in the vicinity of a near-field fiber probe tip [8] and in lithography where optical near-fields appear in the surroundings of a mask pattern that induces chemical reactions in a photoresist [9].

One unique attribute inherent in optical near-field interactions is that it exhibits a hierarchical nature, meaning that the optical near-fields behave differently depending on the physical scale involved, as schematically shown in Figure 12.1(a).

In this chapter, with a view to exploiting such a hierarchical nature in optical near-fields to nanofabrication, we highlight the process of generating smaller-scale structures from larger-scale ones via optical near-field interactions [11], which is schematically shown in Figure 12.1(b). Usually, the generation of smaller-scale structures involves higher costs in achieving the required fine precision. Instead, if we could produce the intended fine structures with less physically demanding resources, manufacturing costs could be reduced. Moreover, since this is a unique ability of optical near-field interactions, it would be technically difficult to mimic the resultant structure using other fabrication methods; therefore, potential applications would include security-related functions, for example, in so-called physical unclonability [12] and so forth.

Nanophotonics and Nanofabrication. Edited by Motoichi Ohtsu
Copyright © 2009 WILEY-VCH Verlag GmbH & Co. KGaA, Weinheim
ISBN: 978-3-527-32121-6

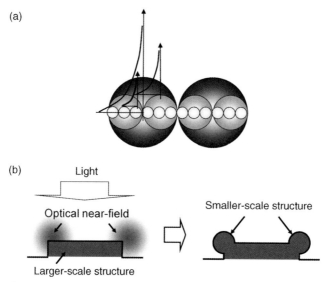

Figure 12.1 (a) Hierarchical nature of optical near-fields. (b) Generation of smaller-scale structures from larger-scale ones via optical near-field interactions.

This chapter is organized as follows. In Section 12.2, we describe an angular spectrum representation of electric fields, which explicitly represents the scale dependence of optical near-fields. With this way of representation, in Section 12.3, we theoretically deal with hierarchical properties in optical near-fields that are exploited for generating smaller-scale structures from larger-scale ones. In Section 12.4, we describe an experimental demonstration of such principles by using ZnO nanoneedles fabricated through metalorganic vapor phase epitaxy (MOVPE) followed by a photoinduced MOVPE procedure, where smaller-scale generated structures were clearly observed with the help of light irradiations. We also observed that the generated fine structures followed a power-law distribution, indicating that fractal structures emerged via optical near-field interactions. Section 12.5 concludes the chapter.

12.2
Angular Spectrum Representation of Optical Near-Fields

An angular spectrum representation of electromagnetic fields [13] is one of the most adequate frameworks to grasp the hierarchical nature in optical near-fields [14]. This approach allows analytical treatment and gives an intuitive picture of the localization of optical near-fields in the region a given distance away from the material since it describes electromagnetic fields as a superposition of evanescent waves with different decay lengths and corresponding spatial frequencies. In other words, one can grasp the fact that an optical near-field at a point of interest originates from a structure whose scale is comparable to the distance between the material and the point of interest [15]. We first briefly introduce the framework of the angular spectrum representation, followed by its actual application to a single-dipole system in order to prepare for the later discussion on hierarchy in optical near-field interactions.

We consider an oscillating electric dipole $\boldsymbol{d}^{(s)}$ with frequency K placed in a vacuum at the origin of a Cartesian space in which the velocity of light is taken as unity ($c=1$). To deal with scattering problems based on assumed planar boundary conditions, it is convenient to represent the complex amplitude $\boldsymbol{E}$ of electric-dipole radiation as a superposition of plane waves with complex wave vector, called the angular spectrum representation [14], defined by

$$\boldsymbol{E}(\boldsymbol{r}) = \left(\frac{iK^3}{8\pi^2\varepsilon_0}\right) \sum_{\mu=TE}^{TM} \int_0^{2\pi} d\beta \int_0^{\infty} ds_{\parallel} \frac{s_{\parallel}}{s_z} \left[\varepsilon(\boldsymbol{s}^{(\pm)},\mu) \cdot \boldsymbol{d}^{(S)}\right] \varepsilon(\boldsymbol{s}^{(\pm)},\mu) \exp(iK\boldsymbol{s}^{(\pm)} \cdot \boldsymbol{r})$$

(12.1)

where the unit wavevector and the transverse-electric (TE) and transverse-magnetic (TM) polarization vectors are given, respectively, by

$$\boldsymbol{s}^{(\pm)} = (s_{\parallel}\cos\beta, s_{\parallel}\sin\beta, \pm s_z).$$

$$\varepsilon(\boldsymbol{s}^{(+)}, TE) = (-\sin\beta, \cos\beta, 0) \quad (12.2)$$

$$\varepsilon(\boldsymbol{s}^{(+)}, TM) = (\pm s_z\cos\beta, \pm s_z\sin\beta, -s_{\parallel})$$

and the parameter s_z of the unit wavevector is represented by

$$s_z = \begin{cases} \sqrt{1-s_{\parallel}^2} & \text{for } 0 \le s_{\parallel} < 1 \\ i\sqrt{s_{\parallel}^2-1} & \text{for } 1 \le s_{\parallel} < +\infty. \end{cases} \quad (12.3)$$

We denote the complex wavevector as $\boldsymbol{s}^{(+)}$ for $z>0$ and $\boldsymbol{s}^{(-)}$ for $z<0$. One of the notable features of the angular spectrum representation is that the value of $s_{\parallel}$ specifies the nature of a plane wave as a homogeneous wave ($0 \le s_{\parallel} < 1$) or an evanescent wave ($1 \le s_{\parallel} < +\infty$). Since we consider optical interactions of the subwavelength regime where evanescent components are dominant and homogeneous ones are negligible, we focus in the region of $1 \le s_{\parallel} < +\infty$ where $s_{\parallel}$ indicates the spatial frequency of an evanescent wave propagating in the assumed planar boundary plane, namely the xy-plane.

Now, we analyze the hierarchy using the angular spectrum representation. Suppose, for example, that we have an oscillating electric dipole $\boldsymbol{d}^{(k)}$ on the xz-plane, where k specifies each dipole. It is oriented at an angle of $\theta^{(k)}$ with respect to the z-axis and at an angle of $\phi^{(k)}$ in the xy-plane,

$$\boldsymbol{d}^{(k)} = d^{(k)}(\sin\theta^{(k)}\cos\phi^{(k)}, \sin\theta^{(k)}\sin\phi^{(k)}, \cos\theta^{(k)}) \quad (12.4)$$

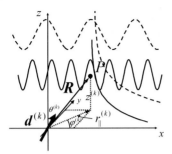

Figure 12.2 Geometrical arrangement of a dipole and a point of interest for an angular representation of optical near-fields.

as schematically shown in Figure 12.2. Suppose also that we observe the radiation from the electric dipole $\boldsymbol{d}^{(k)}$ at a position displaced from the dipole by

$$\boldsymbol{r}^{(k)} = \left(r_{\parallel}^{(k)} \cos\varphi^{(k)}, r_{\parallel}^{(k)} \sin\varphi^{(k)}, z^{(k)} \right) \tag{12.5}$$

The angular spectrum representation of the z component of the electric field for evanescent waves (that is, $1 \leq s_{\parallel} < +\infty$) from dipoles $\boldsymbol{d}^{(k)}$ ($k = 1, \cdots, N$), where N is the number of dipoles, is given by

$$E_z(\boldsymbol{r}) = \left(\frac{iK^3}{4\pi\varepsilon_0} \right) \int_1^{\infty} ds_{\parallel} \frac{s_{\parallel}}{s_z} f_z\left(s_{\parallel}, \boldsymbol{d}^{(1)}, \cdots, \boldsymbol{d}^{(N)} \right) \tag{12.6}$$

where

$$\begin{aligned} &f_z\left(s_{\parallel}, \boldsymbol{d}^{(1)}, \cdots, \boldsymbol{d}^{(N)} \right) \\ &= \sum_{k=1}^{N} \left\{ d^{(k)} s_{\parallel} \sqrt{s_{\parallel}^2 - 1} \sin\theta^{(k)} \cos\left(\phi^{(k)} - \varphi^{(k)} \right) J_1\left(Kr_{\parallel}^{(k)} s_{\parallel} \right) \exp\left(-Kz^{(k)} \sqrt{s_{\parallel}^2 - 1} \right) \right. \\ &\left. + d^{(k)} s_{\parallel}^2 \cos\theta^{(k)} J_0\left(Kr_{\parallel}^{(k)} s_{\parallel} \right) \exp\left(-Kz^{(k)} \sqrt{s_{\parallel}^2 - 1} \right) \right\} \end{aligned} \tag{12.7}$$

Here, $J_n(x)$ represents a Bessel function of the first kind. Here, $f_z(s_{\parallel}, \boldsymbol{d}^{(1)}, \cdots, \boldsymbol{d}^{(1)})$, which we call the angular spectrum of the electric field, describes how components with higher spatial frequency decays depending on the distance from the sources [14]. Therefore, we can analytically and intuitively obtain the insight regarding the localization of electric fields depending on the point of observation, such as at the position $\boldsymbol{r}$ in Figure 12.2. Also, as shown in (12.7), it explicitly contains physical parameters associated with the orientations of the electric dipoles. Therefore, the angular spectrum representation is well suited for analyzing the hierarchical nature of optical near-fields.

12.3
Generation of Smaller-Scale Structures via Optical Near-Fields: A Theoretical Basis

Now, we present a physical model of structures based on multiple dipoles. Suppose that there is a structure whose size is represented by a horizontal length denoted by

12.3 Generation of Smaller-Scale Structures via Optical Near-Fields: A Theoretical Basis

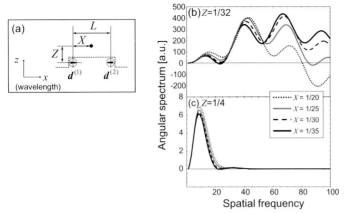

Figure 12.3 (a) Physical model based on dipoles ($d^{(1)}$ and $d^{(2)}$) and the points of interest denoted by X and Z. The size of the larger-scale structure is represented by L. Angular spectrum evaluated at positions (b) closer to the structure ($Z=1/32$), and (c) relatively far from the structure ($Z=1/4$). The dimensions are represented in units of wavelength.

L, as shown by the dashed lines in Figure 12.3(a). When we irradiate this structure with light, as described in Section 12.1, electron charges tend to be concentrated at the corners of the structure. Therefore, as a phenomenological model, we represent such a situation by a two-dipole model, as shown in Figure 12.3(a), where the dipoles are labeled $d^{(1)}$ and $d^{(2)}$. We also assume that those dipoles have a phase difference of π and are parallel to the x-axis; therefore we assume that the orientations of the dipoles are given by $\theta^{(1)} = -\pi/2$ and $\theta^{(2)} = \pi/2$. The distance between the dipoles is given by L.

In order to show the scale dependence of optical near-fields in this model, we consider the electric field at a position that is away from the dipole $d^{(1)}$ by distances X and Z along the x-axis and z-axis, respectively, as shown in Figure 12.3(a). Here, L, X, and Z indicate distances in units of wavelength. If the angular spectrum contains higher spatial frequency components, which means that the electric field is localized at that position to the extent given by that spatial frequency.

Here, we assume that the size of the structure is $L=1/4$. In order to evaluate localization of optical near-fields in the vicinity of the structure denoted by L, first we analyze the region close to the structure, namely $Z=1/32$. Figure 12.3(b) represents the angular spectrum at different horizontal distances $X=1/20$, $1/25$, $1/30$, and $1/35$. We found that at smaller X, namely, when X is smaller than $1/30$, the angular spectrum has positive values in the spatial frequency range denoted by $1 < s_\| < 100$. This indicates that electric fields are strongly localized in the region close to $d^{(1)}$, which corresponds to the corner of the large structure denoted by L. As the point of interest approaches the midpoint between the dipoles, the angular spectrum has negative values, for example, with the cases $X=1/20$ and

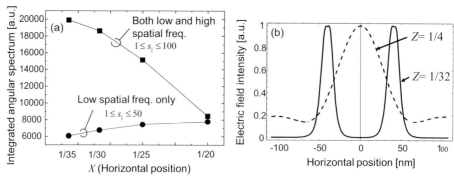

Figure 12.4 (a) Integrated angular spectrum in the spatial frequency interval of $1 \leq s_{\parallel} \leq 100$ and $1 \leq s_{\parallel} \leq 50$ at $Z = 1/32$. (b) Electric-field intensity profiles at $Z = 1/32$ (solid curve) and $Z = 1/4$ (dashed curve) calculated through rigorous theoretical modeling.

1/25 in Figure 12.3(b), meaning that the localization of the electric field becomes relatively degraded as we get closer to the midpoint between the dipoles, or the midpoint of the large structure denoted by L. In other words, a localized electric field appears in the vicinity of the corner of the structure; a smaller scale structure represented by $\Delta X = 1/30$ appears from a relatively large structure denoted by $L = 1/4$.

Figure 12.4(a) also demonstrates such a mechanism clearly by integrating the angular spectrum over lower and higher spatial frequencies, while Z is kept at 1/32. We can see that the integral of the angular spectrum over the domain $1 \leq s_{\parallel} \leq 100$, indicated by squares, increases as the horizontal position gets closer to the corner of the structure. On the other hand, the integral of the angular spectrum over the domain $1 \leq s_{\parallel} \leq 50$, represented by circles, exhibits nearly the same value regardless of the horizontal position. These results demonstrate that the high spatial-frequency components contribute to the localization of electric fields in the vicinity of the corner of the structure.

In contrast, at a relatively large distance away from the structure, for instance $Z = 1/4$, the angular spectrum exhibits nearly the same distribution regardless of the value of X, as shown in Figure 12.3(c). This means that the optical near-fields are uniformly distributed at this scale; a smaller-scale structure is not generated at regions far from the large structure denoted by L.

In fact, we can derive the electric-field distribution at any given point based on a rigorous representation of the electric field, including far-field and near-field components. The solid and dashed curves in Figure 12.4(b), respectively, show the intensity profiles along the x-axis at distances far from the dipoles, namely, distances $Z = 1/32$ and $1/4$, respectively. Here, we assume a wavelength of 325 nm. We can clearly observe that localization appears strongly in the vicinity of the material ($Z = 1/32$), but does not appear relatively far from the material ($Z = 1/4$).

12.4
Experiment

To demonstrate such scale-generation processes via optical near-field interactions, we used ZnO nanoneedles as a generator of optical near-fields. The metallic nanostructures were deposited on the ZnO nanoneedles by a photoinduced chemical reaction. The effect of light irradiation was evaluated by comparing the resultant structures grown in the areas where light was irradiated and where it was not.

ZnO nanoneedles were grown on a sapphire (0001) substrate using a metalorganic vapor phase epitaxy (MOVPE) system [16]. Diethyl zinc (Et_2Zn) and oxygen were used as reactants, and they had flow rates ranging from 0.5 to 5 sccm and 20 to 100 sccm, respectively. The deposition temperature was 400 °C, and the growth time was 1 h.

The surface morphology of the as-grown ZnO nanoneedles was investigated using scanning electron microscopy (SEM). As shown in Figure 12.5(a), a high density of ZnO nanoneedles was vertically aligned over the entire substrate, and they exhibited sharp tips. Typically, the nanoneedles exhibited a mean tip radius of 5 nm, which is expected to generate strong optical near-fields.

After growing the nanoneedles, we deposited Zn nanoparticles on the ZnO nanoneedles. To realize near-field deposition [17], instead of thermal deposition, we conducted deposition using a photoinduced chemical reaction at low temperature, so that thermal deposition was negligible. Since the Et_2Zn is dissociated at

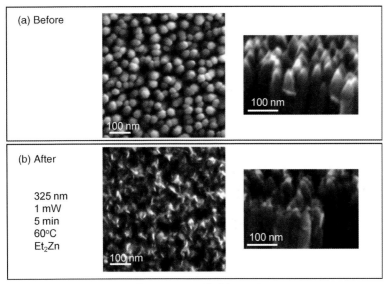

Figure 12.5 (a) SEM images of ZnO nanoneedle fabricated via metalorganic vapor phase epitaxy (MOVPE). (b) SEM images of the same ZnO nanoneedles after photoinduced MOVPE, namely after Zn deposition with light illumination on the ZnO nanoneedles.

temperatures exceeding 150 °C [18], we deposited Zn at 60 °C and used a He-Cd laser with a wavelength of 325 nm as a light source for dissociating the Et_2Zn. The flow rates of Et_2Zn including Ar carrier gas ranged from 20 to 100 sccm. We irradiated He-Cd laser light with an average power of 1 mW for 5 min.

Figure 12.5(b) show SEM pictures of nanoneedles after the Zn deposition processes using light illumination. As is clearly indicated in Figure 12.5(b), fine structures appeared at the apex and vertex of the nanoneedles. We attributed the generation of these fine structures to the optical near-field interactions induced by the nanoneedles, which accelerated the deposition rate of Zn nanoparticles.

We then numerically demonstrated the effect of smaller-scale generation. We analyzed the SEM images to determine the representative scales; these are highlighted as shown in Figure 12.6(a) and (b), which were, respectively, obtained by digitization of images shown in the left-hand side of Figure 12.5(a) and (b). Physically, Figure 12.6(a) represents the projected area of the ZnO nanoneedles, and Figure 12.6(b) shows that of deposited Zn on top of the ZnO nanoneedles. We evaluated the representative scales, denoted by S, by the horizontal extent of the

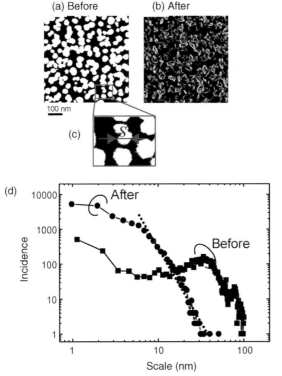

Figure 12.6 Analysis of scales in the structures before and after the photoinduced MOVPE. (a) and (b) Digitized images of the structures. (c) The horizontal extent of the structure used for analysis of the scales denoted by S. (d) Incidence of the scales in (a) and (b).

structures, as schematically shown in Figure 12.6(c). The incidence of these scales S was measured from all rows in the images. Figure 12.6(a) and (b) are, respectively, composed of 510×510 pixels and 591×591 pixels, but occupy the same area of 576 nm × 576 nm.

Figure 12.6(d) shows the incidence of the scales S, where the square and circular marks indicate the structures before and after the Zn deposition with light irradiation, respectively. The original structures exhibited a maximum incidence around 30 nm, which refers to the representative scale S of this structure. On the other hand, the structures after the deposition processes with light exhibited a quite different incidence pattern with smaller scales and higher populations. In other words, smaller-scale structures were generated from larger-scale ones though the light-irradiation process, which induced optical near-field interactions.

Furthermore, it should also be noted that the histogram of the structures fabricated with light exhibits a power-law distribution; the square marks were fitted to a straight line in a double logarithmic plot at scales larger than around 6 nm. The incidence of the scale S follows $y = 5 \times 10^6 \, S^{-4.24}$, denoted by a dotted line in Figure 12.6(d), where y represents the incidence. This means that a fractal nature emerged from a nonfractal structure in the scale merit defined above. We are now studying the origin of this fractal behavior based on analysis of the geometries of the ZnO nanoneedles and the dynamics involved in optical near-fields appearing in the vicinity of nanoneedles, as well as their associated Zn deposition processes.

12.5 Conclusion

We have demonstrated the generation of smaller-scale structures from larger-scale structures via optical near-field interactions. A theoretical framework is shown based on an angular spectrum representation of optical near-fields, which allows us to deal explicitly with their hierarchical properties. Experimental demonstrations are also shown using ZnO nanoneedles fabricated through standard MOVPE and Zn nanoparticles deposited through photoinduced MOVPE processes. The incidence of the representative scale clearly shifted to smaller values in the presence of light irradiation. The emergence of fractal behavior was also observed. We are now pursuing further analysis of such small-scale generation processes both theoretically and experimentally. We are also seeking potential applications of these unique physical processes involving optical near-field interactions, such as security applications.

Acknowledgments

The authors are indebted to Prof. Gyu-Chul Yi of Pohang University of Science and Technology (POSTECH), Pohang, Korea for their support in fabricating the nanoneedles.

References

1 Ohtsu, M., Kobayashi, K., Kawazoe, T., Yatsui, T. and Naruse, M. (2008) *Principles of Nanophotonics*, Taylor and Francis, Boca Raton.

2 For example, Yatsui, T., Kourogi, M., Tsutsui, K., Takahashi, J. and Ohtsu, M. (2000) High-density-speed optical near-field recording-reading with a pyramidal silicon probe on a contact slider. *Optics Letters*, **25**, 1279–1281.

3 Chen, S.-J., Chien, F.C., Lin, G.Y. and Lee, K.C. (2004) Enhancement of the resolution of surface plasmon resonance biosensors by control of the size and distribution of nanoparticles. *Optics Letters*, **29**, 1390–1392.

4 Kawazoe, T., Kobayashi, K., Sangu, S. and Ohtsu, M. (2003) Demonstration of a nanophotonic switching operation by optical near-field energy transfer. *Applied Physics Letters*, **82**, 2957–2959.

5 Naruse, M., Miyazaki, T., Kawazoe, T., Kobayashi, K., Sangu, S., Kubota, F. and Ohtsu, M. (2005) Nanophotonic Computing Based on Optical Near-Field Interactions between Quantum Dots. *IEICE Transactions on Electronics*, **E88-C**, 1817–1823.

6 Matsuda, K., Saiki, T., Nomura, S., Mihara, M., Aoyagi, Y., Nair, S. and Takagahara, T. (2003) Near-Field Optical Mapping of Exciton Wave Functions in a GaAs Quantum Dot. *Physical Review Letters*, **91**, 177401 1–4.

7 Yatsui, T., Yi, G.-C. and Ohtsu, M. (2006) Integration and evaluation of nanophotonic device, in: *Progress in Nano-Electro-Optics V* (ed M. Ohtsu), Springer, Berlin, pp. 63–107.

8 Yatsui, T., Takubo, S., Lim, J., Nomura, W., Kourogi, M. and Ohtsu, M. (2003) Regulating the size and position of deposited Zn nanoparticles by optical near-field desorption using size-dependent resonance. *Applied Physics Letters*, **83**, 1716–1718.

9 Yonemitsu, H., Kawazoe, T., Kobayashi, K. and Ohtsu, M. (2007) Nonadiabatic photochemical reaction and application to photolithography. *Journal of Luminescence*, **122–123**, 230–233.

10 Yatsui, T., Nomura, W. and Ohtsu, M. (2005) Self-assembly of size- and position-controlled ultralong nanodot chains using near-field optical desorption. *Nano Letters*, **5**, 2548–2551.

11 Naruse, M., Yatsui, T., Hori, H., Kitamura, K. and Ohtsu, M. (2007) Generating small-scale structures from large-scale ones via optical near-field interactions. *Optics Express*, **15**, 11790–11797.

12 Skoric, B., Maubach, S., Kevenaar, T. and Tuyls, P. (2006) Information-theoretic analysis of capacitive physical unclonable functions. *Journal of Applied Physics*, **100**, 024902 1–11.

13 Wolf, E. and Nieto-Vesperinas, M. (1985) Analyticity of the angular spectrum amplitude of scattered fields and some of its consequences. *Journal of the Optical Society of America A-Optics Image Science and Vision*, **2**, 886–890.

14 Inoue, T. and Hori, H. (2005) Quantum theory of radiation in optical near-field based on quantization of evanescent electromagnetic waves using detector mode, in: *Progress in Nano-Electro-Optics IV* (ed M. Ohtsu), Springer, Verlag, pp. 127–199.

15 Naruse, M., Inoue, T. and Hori, H. (2007) Analysis and Synthesis of Hierarchy in Optical Near-Field Interactions at the Nanoscale Based on Angular Spectrum. *Japanese Journal of Applied Physics*, **46** (9A), 6095–6103.

16 Park, W.I., Yi, G.-C., Kim, M. and Pennycook, S.J. (2002) ZnO nanoneedles grown vertically on Si substrates by non-catalytic vapor-phase epitaxy. *Advanced Materials*, **14**, 1841–1843.

17 Yatsui, T., Kawazoe, T., Ueda, M., Yamamoto, Y., Kourogi, M. and Ohtsu, M.

(2002) Fabrication of nanometric single zinc and zinc oxide dots by the selective photodissociation of adsorption-phase diethylzinc using a nonresonant optical near-field. *Applied Physics Letters*, **81**, 3651–3653.

18 Kuniya, Y., Deguchi, Y. and Ichida, M. (1991) Physicochemical properties of dimethylzinc, dimethylcadmium and diethylzinc. *Applied Organometallic Chemistry*, **5**, 337–348.

Index

a
absorption coefficient 219
acrydite-DNA sequence 228
acrydite-terminated DNA 228
adiabatic photolithography 28
– schematic presentation 28
adiabatic process 10, 11, 29
– fabrication 11
– photoresist 29
Al-coated mold 57, 59
AlGaAs barrier layer 76
angular frequency 24
angular spectrum 12, 245, 247, 248
annealing process 95, 97
antimonide-related material 80, 90
aperture pitch 135
APTES-guided technique 233
– molecular self-assembly 232
argon-ion laser 197
artificially assisted self-assembling (AASA) 173
atom-detecting devices 12
atom photonics 12
atomic-force microscope (AFM) 28, 29, 47, 58, 61, 72, 96
– image 28, 32, 47, 58, 62, 75, 76, 93
atomic-level material fabrication 2
– sapphire 37

b
band-edge emission 112
bandgap semiconductors 71
band-pass filters 28
bar magnets 168
biotin-cDNA chains 228
bit-patterned media (BPM) 167
block-copolymer films 153, 159
– lithography 154

Bragg condition 182
bulk materials 116

c
calculation models 56
– schematic of 56
capillary force 49, 51
carboxyl group 51, 54
Carniglia–Mandel model 12
catalyst-assisted vapor-liquid-solid (VLS) 107
catalyst-free metalorganic vapor-phase epitaxy, 107, 108
charge-neutrality equation 218
chemical-mechanical polishing (CMP) 60
chemical vapor deposition (CVD) 36, 243
– photodissociation 17
– processes 243
chip-to-chip interconnects 215
chlorine-based reactive ion etching (RIE) 95
coherent magnetization rotation 158
colloidal beads 50
colloidal gold nanoparticles 49, 51, 52, 53, 54
– self-assembly 49
– SEM image 52
complementary DNA (cDNA) 226
– biotin-linked 226
cone angle 20

d
DC magnetron sputtering methods 155
Debye–Waller factors 180
density/emission wavelength 74
deposition rate 26
– higher-order power dependence 26
deposition techniques 45
diamond-like carbon (DLC), 211

– film surface   195, 197, 200, 202, 204, 208, 209, 210, 212
– Raman spectra   197
– stripes   205
dielectric wave guides   235
diffraction efficiency curves   186, 187
dipole–forbidden transition   12
DNA-directed self-assembly   227
– mediated methods   216
– programmable   227
dot arrays   159
DRAM technology   35
dry-etching process   150

e
EB-patterned GaAs surface   95
electric dipole operator   24
– components   24
electric-field intensity distribution   48
electroluminescence devices   102
electromagnetic field   1, 4, 5, 134, 135, 185, 244
– distribution of   134
– spectrum representation   244
electromagnetic interaction   4
electron-beam lithography   92, 95, 137, 152, 170, 224
– equipment   71
electron-beam writing technology   9, 32, 180
electron-hole pair   216
electron microscopy images   114
electron orbital energies   179
electron projection lithography (EPL)   55
electronic devices   91
electronic dipoles   3
electronic integrated circuits   100
– technologies   215
electronic/photonic devices   107
energy-beam irradiation   150
energy-density excitation   193
energy-dispersive X-ray spectroscopy   117
energy gap   71
excitonphonon-polariton (EPP)   20, 61
– energy   24
– model   20, 24, 26
exposure system   136

f
fabricating arbitrary nanostructures   71
fabrication methods   232
fabrication processes   150, 151, 216
fabrication/quantitative innovations   36
fabrication technique   73, 100, 101
fabrication technology   71, 89, 145

far-field light   20
far-field propagating light   20
FePt dots   156, 158
– array   156, 162
fiber-optic communication networks   83
fiber probe   10, 20, 44
field-emission gun scanning electron microscopy (FEG-SEM)   109
finite-difference time-domain (FDTD)   134, 216
– calculation   58
– method   30, 55, 58, 170, 221
– simulation   30, 217, 220, 236
free electrons   211
full width at half-maximum (FWHM)   78, 187

g
GaAs buffer layer   74, 80, 93, 95, 96, 98
GaAs spacer layer   97
GaAs substrate   72, 80, 90, 95
gain-dependent QD waveguide transmission   223
– Monte Carlo simulation   223
GaN micropyramids   127
gas-phase DEZn   37, 41
gas-phase diethylzinc   17
GaSb surface   93
GC spectrum   204
GC surface   205
geometrical arrangement   246
gold ions   51
– reduction of   51
gold nanoparticles   53
growth mode transition   112

h
hard disk drives   131, 147
high-density InGaSb QD layer   82
high-density magnetic recording   168
high-density QDs   78
– self-assembled QDs   83
high-density storage system   167
high-performance devices   73
– photodetectors   73, 78
– semiconductor laser   70
high-precision material   194
high-quality nanorod heterostructures   108
high-resolution optical lithography   224
high-resolution processing   193
high-speed electronic devices   105
high-speed signal processing   73
high quality fiber probes   2
higher-magnification SEM images   51

hydride vapor-phase epitaxy   107
hydrofluoric acid solution   102

i

InAs   79
– AFM image   78, 79, 80
– QDs   78, 80, 90
InGaAlAs spacer layers   85
InGaAlAs strain-compensation layers   85
InGaSb QDs   81, 82
– AFM image   82
– high-density   81
in situ vacuum shear-force microscopy   36
integrated angular spectrum   248
interband transition state   112
interdot coupling efficiency   221

k

KAP crystal   187

l

laminar-type multilayer grating   181
– schematic diagram   181
laser pulse energy   198
lattice constant   85, 90
lattice-matched InAlAs buffer layer   85
lattice-mismatched material system   71
lattice-mismatched semiconductor material systems   80
light-emitting devices   105
light-emitting diodes (LEDs)   105, 106
– arrays   127
light-emitting semiconductor particles   9
– GaN   9
– ZnO   9
light absorption/emission   69
light-matter interactions   12, 193
lithography system   131
lithography technology   150
low-density quantum dots   73, 76, 78
low-intensity propagating light   11
low-noise high-speed responses   83

m

macroscopic subsystem   4
magnetic crystal anisotropy energy   147
magnetic domain   168
magnetic dots   149, 150
magnetic field   148
magnetic force microscopy (MFM)   172
– image   172, 174
magnetic hysteresis curves   157
magnetic islands   158, 167
magnetic materials   167

magnetic media   148
– schematic explanation   148
magnetic nanodots   155, 175
– fabrication of   155
magnetic recording   168, 177
material systems   85
Maxwell's equations   185, 220
metal catalyst-assisted vapor-liquid-solid (VLS)   107
metal nanoparticle   41, 215
metallic nanodot   46
metallic nanoparticles   45, 46, 48
metallic waveguides   3
metalorganic molecule   10
metalorganic vapor phase epitaxy (MOVPE)   36, 244
– catalyst-free   107
– system   107, 108
– technique   107
microfabrication processes   107
Mie's theory of scattering   42, 48
molecular beam epitaxy (MBE)   71, 92, 107
– growth chamber   85
– method   96
molecular electronic/vibrational energies   25
molecular-flux intensity   99
molecular vibration modes   23
monochromator crystals   179, 180
Monte Carlo approach   236
multiphoton absorption process   22
multiple QD-type waveguide fabrication process   231
multipolar quantum electrodynamics   24
multiquantum wells (MQWs)   92, 107
– Sb-QDs   92

n

nanodot chain   48
nanofabrication technologies   215
– advances   215
– schematic drawing   217
nanoimprint lithography   132, 151
– process   152
– templates   152
nanojet probe method   96
nanometer-scale   116
– dots   39, 40
– photonic   105
nanometer-sized templates   49
nanometric metallic particles   1, 4, 5, 9
– Al   9
– Zn   9
nanopatterned media   149, 160
– schematic explanation   149

nanophotonic devices 2, 8, 9, 49, 70, 73, 75, 81, 83, 105, 106, 126
– design 2
– operation 8
nanophotonic fabrication 35
nanophotonic/macroscopic photonic devices 7
nanophotonic switch 7, 8, 11
nanophotonic systems 11
– quantum-dot system 4, 12
nanophotonics technology 5, 243
nanopositioned AlSb/GaSb QW structure 93
nanopositioned Sb-QW structure 93
nanorod axial MQWs 109
nanorod heterostructures 105, 106, 107, 122
– schematics 106
nanorod quantum structures 105, 106, 107, 112, 124, 127
nanorod SQW structures 115, 116
– schematic 115
nanoscale ablation 206
nanoscale structures 48
– corrugations 211
nanosized QW structures 126
nanostructure devices 132, 211
nanostructure fabrication 60
nanostructured Sb-based semiconductors 100
near-field desorption 45
near-field etching 63
near-field generator 167
near-field intensity distribution 220
near-field lithography 29, 132, 137
– optical 137
– lithography 55, 59
near-field microscopy 91
near-field optical chemical vapor deposition (NFO-CVD) 18, 26, 35, 36, 44
– nonadiabatic 26
– schematic explanation 18, 36
– shear-force images 26
near-field optics 1, 2
near-field phenomena 35
near-field photoluminescence spectrometers 2
near-field photomask 133, 136, 142, 143
near-field scanning optical microscopy (NSOM) 124
near-field spectrometer 2
negative-dielectric materials 236
negative-index optical fibers 215
next-generation lithography (NGL) 55
nonadiabatic optical near-field etching 62
nonadiabatic photochemical process 27, 29, 31
nonadiabatic photochemical reaction 35, 60, 184
nonadiabatic photodissociation process 26
nonadiabatic photolithography 27, 28
– schematic presentation 28
nonadiabatic process 10
nonadiabatic lithography 32
– AFM imagesof 32
nonmagnetic matrix 147
nonradiative competitive process 218
nonresonant optical near-field 19
nonresonant process 5

o

ONF intensity 20
optical communication devices 72
optical-gain coefficient 222
optical-gain distributions 91
optical-gain materials 93
optical fiber probe 36
optical field intensity 57
– localization of 57
optical nanofabrication techniques 9
optical near-field 145
– interactions 243
– lithography 138, 140, 144, 145
– use 3
optical/magnetic hybrid disk storage density 11
optical power density 28, 29
opto electronic devices 105
– nanodevices 105
organo-silane molecule 224

p

particle-suspension contact angle 50
PbS photodetector 73
periodic ablation 201
phase-shift mask technology 136
phonon-coupled devices 7
– nanophotonic switches 7
photochemical etching method 102
photochemical reaction 9, 10, 19, 60
– adiabatic 60
photochemical vapor 10
– deposition 9
photochemically etched layers 102
– emission spectra 102
photodeprotection resist system 142
photodissociation 38
– process 22
– reaction 44

– schematic diagrams 38
photoelectron spectroscopy 22
photolithography 9
– technology 145
photoluminescence spectrum 94
photomask 27
– schematic diagrams 27
photomultiplier tube (PMT) 73
photon energy 9, 36, 43, 46, 48, 60, 111, 113, 123, 179
photonic devices 1, 5, 70, 83, 93, 105
– nanophotonic switches 105
photonic integrated circuits 215
photonic/electronic devices 27
photonic systems 5
PL emission energies 112
PL spectra 75, 78, 113, 122, 124
– measurement system 73
plasma oscillation 9
PMMA layer 232
polishing method 63
polishing pad 63
polishing particles 60
polymeric resin 198
polystyrene-poly-methylmethacrylate (PMMA) 172
position-controlled dot-chain formation 48
position-controlled nanoscale structures 48
potential curves 23
– schematic diagram 23
processor chip speeds 215
propagating light-coupled device 7
propylene glycol monomethyl ether acetate (PGMEA) 155
prototype device 138, 144
pulse energy 199
pump-probe experiment 204, 212
– technique 202
pump pulse 202

q

quantum dots 6, 91, 92
– array 88, 217, 222, 234
– based nanophotonic circuits 228
– density 75
– devices 99
– diameter 77, 222
– emission spectrum 75, 226
– fabrication process 225
– fabrication technique 71, 73
– formation 88
– layers 75, 108
– materials 76
– photonic devices 216

– semiconductor optical amplifier (SOA) 73
– structures 91, 93, 100, 116
– waveguide 216, 222, 224, 232, 235, 236
quartz glass slide 174
quasi-Fermi energies 218
– utilization 116

r

radial nanorod heterostructures 107, 123
radial nanorod quantum structures 105
radial nanorod SQWs 122, 125
Raman spectroscopy 197, 198, 199, 201
RCWA method 185, 189
reactant-gas delivery system 107
refractive index 55, 58, 181, 194
rigorous coupled wave analysis (RCWA) 185
– method 185

s

sapphire substrate 21
– shear-force image 21
Sb-based QD 92, 100, 101
– AFM image 93
– semiconductors 100
– shatter sequence 101
– structures 92
Sb-flux intensity 100
scanning electron microscope (SEM) image 58, 170
scanning near-field optical microscope (SNOM) 73
scanning transmission electron microscope (STEM) 85
scanning tunneling microscope (STM) probe 98
selective-area formation technique 102
selective-area growth technique 99
self-assembled monolayer 224
self-assembled QDs 71, 77, 90, 91
– density/wavelength 77, 91
– fabrication 80, 230
self-assembling method 48
semiconductor crystal growth 71
semiconductor devices 89
semiconductor films 71
semiconductor industry 150, 151
semiconductor lasers 71
– realization 71
semiconductor laser diode 73
semiconductor layers 216
semiconductor memory 131
– solid-state drives (SSD) 131
semiconductor nanorod growth methods 108

semiconductor nanorod heterostructures 107
semiconductor nanorod quantum structures 107
semiconductor nanorods 107
semiconductor nanostructures 105
– one-dimensional (1D) 105
semiconductor quantum dot (QD) 69, 215
– structure 100
– surfaces 92
sensing systems 2
Si-doped GaAs thin films 81
Si-MOS compatible nanophotonics technology 100
Si nanoparticles 102
– photochemical etching method 102
Si photodetectors 78
Si wedge structure 53
– SEM image of 53
signal-to-noise ratio (SNR) 147
signal-to-noise ratio 167
signal magnitude 6
silicon-related quantum structure fabrication technology 100
silicon nanoparticles 101
silicon quantum dots 100
silicon waveguides 3
single-crystalline semiconductor nanorods 107
single-electron transistors 105
single quantum-well structures (SQWs) 107
site-controlled fabrication technique 91
site-controlled InAs/GaAs QDs 97
– AFM image 97
site-controlled individual InAs QD structures 98
site-controlled nanostructures 91
– fabrication techniques 91
site-controlled QD 99
size-controlled deposition methods 49
size-controlled dot-chain formation 48
size-dependent resonance 35, 39, 40, 41
smaller-scale structures 244
SNOM measurement 75, 81
solid-state drives (SSD) 131
solid-vapor 198
spin on glass (SOG) 138
SPM images 210
stacked structure 86
– STEM image 86
STEM measurement 87, 88
step-and-repeat method 140
STM-probe assisted nanolithography technique 99

STM-probe-assisted site-controlling technique 98
storage technology 147
strain compensation 84
– layer 85
– method 84
– schematic diagram 84
strain energy 85
streptavidin-bound QDs 226
subdiffraction QD waveguide 217
surface plasmon-polaritons (SPPs) 208
synthetic silica substrates 61

t

temperature-dependent PL spectra 125
thermal annealing process 154
thermal diffusion equation 176
thermally assisted magnetic recording (TAR) 167, 169
– conceptual diagram 169
thermoplastic elastomers 153
thin-film deposition technology 71
time-domain method 29
TiN film 196
– SEM images 196
transmission electron microscopy 109
– images 109, 110
– measurements 112
tri-propylene-glycol-diacrylate monomer 57
two-dimensional fast Fourier-transform (2DFFT) image 88

u

ultraflat silica surface 63
ultraflat surface substrate 60
ultrahigh-density QDs 83, 90
– fabrication 83
ultrahigh-resolution images 2
ultralong nanodot chain 46
ultralow growth rate 79
– shutter sequence 79
ultrathin-film photoresist 144
ultraviolet light 9
– fiber probe 17
U-shaped groove guides 159
UV fiber probe 36
UV-sharpened fiber probe 37

v

vapor pressure 75

w

waveguide-characterization procedure 235
wavelength of self-assembled QDs 73

wave-optical picture 9
wide-bandgap materials 107
– GaN 107
– ZnO 107

x

X-ray absorption spectroscopy 179
X-ray diffraction 109
– gratings 179
X-ray diffractometer 184
X-ray emission spectroscopy 179
X-ray plasma sources 179
X-ray spectrometry 179

z

Zn nanodots 41
– SEM images 41
Zn nanoparticles 43
– radius distributions 43

ZnO 121
– PL spectra 121
ZnO axial nanorod quantum structures 108, 126
ZnO-based nanomaterials 126
ZnO/GaN nanorod heterostructures 126
– quantum phenomena 110
ZnO nanoneedles 249
ZnO nanorods 107, 108, 111, 118, 119, 120, 121
– arrays 107, 127
– MQWs 108
– PL spectra 118
– synthesis methods 127
ZnO QW layer 112
– schematic diagrams 112
ZnO/ZnMgO nanorod 112, 113
– QD formation 113
– TEM image 112